Lecture Notes in Control and Information Sciences

Edited by M. Thoma and A. Wyner

For information about Vols. 1–61 please contact your bookseller or Springer-Verlag.

Lecture Notes in Control and Information Sciences

Edited by M. Thoma and A. Wyner

123

C. W. de Silva
A. G. J. MacFarlane

Knowledge-Based Control with Application to Robots

Springer-Verlag
Berlin Heidelberg GmbH

Authors
Dr. Clarence W. de Silva
NSERC Professor of Industrial Automation
Department of Mechanical Engineering
University of British Columbia
Vancouver, B.C. V6T 1W5
CANADA

Professor Alistair G. J. MacFarlane
Principal and Vice-Chancellor
Heriot-Watt University
Riccarton, Edinburgh EH 14 4AS
SCOTLAND, U.K.

ISBN 978-3-540-51143-4

Library of Congress Cataloging in Publication Data
De Silva, Clarence W.
Knowledge-based control with application to robots / C.W. de Silva, A.G.J. MacFarlane.
(Lecture notes in control and information sciences ; 123)
ISBN 978-3-540-51143-4   ISBN 978-3-540-46168-5 (eBook)
DOI 10.1007/978-3-540-46168-5

1. Robots--Control systems.
2. Intelligent control systems.
3. Expert systems (Computer science).
I. MacFarlane, A. G. J. (Alistair George James). II. Title. III. Series.
TJ 211.35.D42   1989
629.8'92--dc20   89-10088

2161/3020-543210 – Printed on acid-free paper.

PREFACE

The work presented here considers the integration of knowledge-based soft control with hard control algorithms. As a specific application, the development of a knowledge-based controller for robotic manipulators is addressed. Servo control alone is known to be inadequate for nonlinear and high-speed processes including robots. Furthermore, knowledge-based control such as fuzzy control, when directly included in the servo loop, has produced unsatisfactory performance in research robots. These considerations, along with the fact that human experts can very effectively perform tuning functions in process controllers, form the basis for the control structure proposed in this work.

The proposed control structure consists of a three-level hierarchy. The lowest level has a set of PID controllers closed around a high-speed decoupling and linearising controller. A recursive algorithm has been developed for implementation of this nonlinear feedback controller. The second level contains a set of knowledge-based controllers known as servo experts. There is one servo expert for each degree of freedom of the process. A servo expert is a knowledge system of the forward production type. It monitors the corresponding process response and makes inferences on the basis of a set of performance specifications. These inferences are supplied to the fuzzy controller at Level 3. The knowledge base of the fuzzy controller consists of expert knowledge in the form of linguistic rules for servo tuning. These rules are expressed as fuzzy decision tables by using membership functions of the condition and action variables, fuzzy logic, and the compositional rule of inference. This knowledge, along with external sensory data and other available information, is used to arrive at tuning decisions for the PID controllers. Control specifications, parameters of the decoupling controller, and the rule bases themselves can be modified as well, using this type of knowledge, at the fuzzy control level. Separation of the knowledge-based control into two levels, with the lower level functioning as an intelligent preprocessor and the upper level containing a fuzzy knowledge base that is representative of expert knowledge in servo tuning, is an important characteristic of the proposed control structure.

The background work presented includes a thorough literature review of hard algorithmic control and knowledge-based soft control of robotic manipulators. Theory, concepts, and procedures for developing each level in the proposed hierarchical control structure, have been established.

As an application of the proposed control structure, a knowledge-based control system has been developed for a two-degree-of-freedom robot. The system contains a servo expert for each joint of the robot, developed using a commercially available AI toolkit (MUSE). The unstructured code in the servo experts was written in PopTalk language and the structured code was developed using the editor_tool facility of MUSE. The fuzzy controller was developed off line and it has been implemented as a set of decision tables. A robot simulator was developed as a separate UNIX process written in C language. The simulator has been interfaced with the servo experts and with the fuzzy controller using the UNIX Socket facility, and the Channel objects and Data Stream objects that are available with the AI toolkit. Data channel programs were written in PopTalk. Performance of the overall system was evaluted using simulation experiments.

Simulation experiments which were carried out included single-joint tests with step, ramp, and sine inputs. Results have demonstrated the superiority of the proposed control approach in comparison with conventional control without knowledge-based tuning. Simulations of a series of seam tracking tasks were carried out with the complete robot. Initial position errors were introduced and acceleration disturbances were injected during operation, to represent various situations such as pick-and-place operations, fast negotiation of a corner, and operation under an imperfect nonlinear feedback controller. In each simulation, superior results were achieved with the proposed controller, in comparison to a conventional controller. Even though the performance generally improved with the bandwidth of the knowledge-based controller, a bandwidth of at least an order of magnitude smaller than the servo bandwidth consistently provided good performance.The controller was also found to be robust in the sense of having relative insensitivity to initial values, and other parameter values of the controller.

The monograph is divided into seven chapters. Chapter 1 presents the background and the scope of the material given in the remaining six chapters. Chapter 2 reviews the kinematic and kinetic formulation of robots, giving particular consideration to recursive algorithms for computed torque/force control. This material forms the foundation for the recursive algorithm developed in Chapter 4 for linearising and decoupling control of robots, and is an important part in the development of the lowest level of the proposed control structure. Fuzzy logic is used in the development of the top level of the control structure. Chapter 3 gives a simplified review of fuzzy logic and fuzzy control, with geometric illustrations and examples. Chapter 4 presents a general description of the proposed control structure. Chapter 5 describes the application of the knowledge-based control structure to robotic manipulators. The operation of the system developed in Chapter 5 is described in Chapter 6. Simulation experiments are described and representative results are discussed. The computer programs used in this application are listed in Appendix 1. Important features of the proposed control structure are highlighted in Chapter 7, and directions are indicated for further developments. A bibliography of related work is given under References, at the end of the monograph.

This monograph is suitable for students, researchers, and practising professionals in the fields of Automatic Control and Robotics. The material is presented in simple and clear language with sufficient introductory information. Someone with an undergraduate knowledge in dynamics and control should be able to use this book without any difficulty.

Engineering Department
University of Cambridge

C. W. de S and A. G. J. M

ACKNOWLEDGMENTS

We are grateful to Dr. D. J. Williams for encouraging this work and for providing computing facilities. C. W. de Silva wishes to extend his thanks to the Fulbright Commission for the fellowship which, in part, supported this work. The facilities provided by the Engineering Department of the Cambridge University are gratefully acknowledged.

C. W. de Silva, with the endorsement of
Professor A. G. J. MacFarlane,
wishes to dedicate this work to
his mother and to the memory of his father.

TABLE OF CONTENTS

4.   CONTROL STRUCTURE

5.   SYSTEM DEVELOPMENT

1. BACKGROUND

1.1 Introduction

Machine intelligence and robotics are related and complementary fields of study. Computer scientists are active in the development of software which can make computers perform tasks in an apparently intelligent manner. Engineers in the robotics field have achieved satisfactory results in integrating tactile sensing, vision, voice actuation, automatic ranging, and even the sense of smell into robotic systems. In short, there has been a significant effort in making robots more intelligent and human-like. Sensing, actuation, and control are the three key components of robot intelligence. Intelligent control can significantly improve the performance of a robotic manipulator. Intelligence can be incorporated into a controller in the form of a knowledge base, typically expressed as a set of rules, and an associated inference mechanism. The structure of a knowledge-based controller cannot be chosen arbitrarily. The nature of the process, control requirements, and the need for conventional hard controllers have to be properly considered in developing a structure for a knowledge-based control system.

This monograph considers the knowledge-based control of nonlinear and high-bandwidth processes. In particular, a hierarchical control structure which integrates knowledge-based soft control with hard control algorithms is presented. The development of the proposed control structure is centred around its application to the control of robotic manipulators. A robotic manipulator, being a nonlinear and coupled process having a typical operating bandwidth of the order of 100Hz, is a representative process using which the effectiveness of the proposed knowledge-based controller could be evaluated. Robotics, knowledge-based control, and fuzzy logic are central ingredients of the work presented here. This introductory chapter explores those subjects which provide the relevant background.

Since robots are high speed processes which are highly nonlinear, dynamically coupled, and often of high order, it is not adequate only to use linear servo control, if accurate performance in high bandwidth operations is desired. In its normal mode of operation, a robot may be subjected to unplanned events and unfamiliar situations, and it will be required to respond intelligently to these situations. Furthermore, robot performance could be improved if the control system were able to capture and utilise past experience and available human expertise. A knowledge-based control system would be attractive in these respects. But, as some implementations show, it is not advisable to include a knowledge-based controller directly within a servo loop, to generate drive signals for robot joints. Due to the nature of the particular process (a robot), even if the processing speed were not a limitation, a human expert would rely on a hard algorithm for this purpose of control signal generation. Since human experts are quite effective in tuning operations, constantly improving their skills through experience, it is intuitively clear that a knowledge-based controller for a robot may be more effective in a monitoring and tuning capacity than in a direct servo control capacity. A control structure which possesses these intuitively appealing qualities is proposed here.

The proposed control structure has a three-level hierarchy. The lowest level consists of a hard controller containing a group of conventional servo controllers which are closed around a high-speed decoupling and linearising controller. In Level 2 there is a knowledge-based system of the forward production-system type, which consists of an intelligent observer (a servo expert) for each degree of freedom of the robot. The soft controller at the top level of the control structure functions as an intelligent tuner for the hard controller at the lowest level. Fuzzy logic is used for the development of the intelligent tuner. The group of intelligent preprocessors in Level 2 provides inferences regarding a series of attributes of the process response with respect to a set of performance specifications. This information is then supplied to the fuzzy controller in the top level. The fuzzy controller carries knowledge, represented as linguistic statements obtained from human experts, which may containg vague terms, for tuning the joint servos. Based on measurements, inferences from the servo experts, and possibly other types of external information, the fuzzy controller generates tuning commands for the servos at the lowest (hard control) level.

1.2 Objectives

The scope of this monograph is the integration of hard algorithmic control and soft knowledge-based control, in the context of robotic manipulator control. In particular, a hierarchical control structure which has a high-speed hard controller at the lowest level and an intelligent observer and a tuner at the upper levels, is proposed. The monograph explores the rationale for the specific control structure which is proposed here, and develops the necessary qualitative and analytical foundation for the control structure. To demonstrate the development and feasibility of this control structure, an example application has been programmed on a SUN workstation. A two degree-of-freedom robot was simulated as a separate UNIX process using a program written in C language. The servo experts were developed using the MUSE AI toolkit. The top-level fuzzy controller was developed separately using a valid set of linguistic tuning rules for PID servos, and implemented as a set of decision tables within the knowledge-based controller. The robot simulator was interfaced to the servo experts using the UNIX socket facility along with the Channel objects of MUSE and interface programs written in PopTalk language. The fuzzy controller was interfaced with the robot simulator using the Stream objects of MUSE. The control structure was then evaluated on the basis of the simulation results. The merits and shortcomings of the proposed knowledge-based controller are discussed, and directions are indicated for future developments.

1.3 Robot Characterisation

A robot is a mechanical manipulator which can be programmed to perform various physical tasks. Programmability and the associated task flexibility are necessary characteristics for a robot, according to this commonly used definition. Furthermore, a robotic task might be complex to the extent that some degree of intelligence would be required for satisfactory performance of the task. There is an increasing awareness of this (Staugaard, 1987) and there have been calls to include intelligence, which would encompass abilities to perceive, reason, learn and infer from incomplete information, as a requirement in characterising a robot.

A robot can be also interpreted as a control system. Its basic functional components are the structural skeleton of the robot; the actuator system which drives the robot; the sensor system which measures signals for performance monitoring, task learning and playback, and for control (both feedback and feedforward); the signal modification system for functions such as signal conversion (e.g., digital to analog and analog to digital), filtering, amplification, modulation and demodulation;

and the direct digital controller which generates drive signals for the actuator system so as to reduce response error (de Silva, 1985; de Silva, 1988). Higher level tasks such as path planning, activity coordination, and supervisory control are not treated within this basic control system.

Classification

The physical structure of a robot may have anthropomorphic features, but this is a rather narrow perception. Nevertheless there is a classification of industrial robots which is based on kinematic structure. For example, consider the classification shown in Figure 1.1. Six degrees of freedom are required for a robot to arbitrarily position and orient an object in a three-dimensional space. It is customary to assign three of these degrees of freedom to the wrist that manipulates the end effector (hand) and the remaining three to the arm of the robot. Since kinematic decoupling is desired for analytical simplicity (See Chapter 2), spherical wrists having three revolute (R) degrees of freedom with axes of motion coinciding at a single point (at the wrist) are commonly employed. Having decided on this configuration, the kinematic structure of the arm can then be used as a criterion for robot classification.

Specifically, the sequence of rotatory or revolute (R) joints and rectilinear or prismatic (P) joints employed in the arm structure will classify a robot. Four common classifications are shown in Figure 1.1; rectangular or cartesian (3P), cylindrical (R-P-P or P-R-P), spherical or polar (R-R-P), and jointed spherical (3R). Furthermore, SCARA (an acronym standing for selective compliant assembly robot arm) configurations where at least the first two of the three arm degrees of freedom do not face gravity (i.e., vertical revolute axes or horizontal prismatic axes) are desired so that the actuators of the most demanding joints are not subjected to gravity loads.

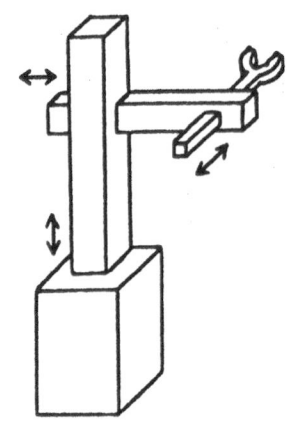

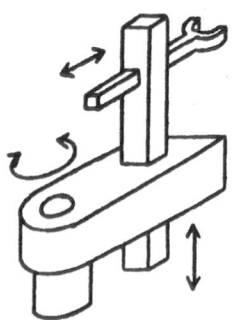

(a) Rectangular (Cartesian)
(3P)

(b) Cylindrical Polar
(R-P-P)

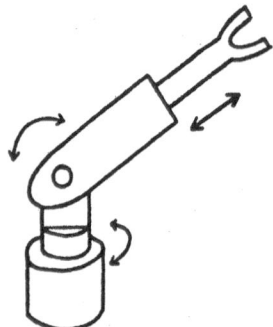

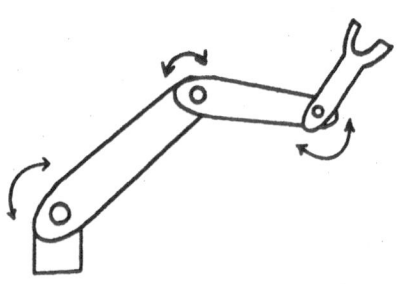

(c) Spherical Polar
(R-R-P)

(d) Jointed Spherical (Articulated)
(3R)

Figure 1.1   A Kinematic Classification for Robotic Manipulators

Other classifications are possible. Robot classification can be made by the actuator type (e.g., hydraulic, DC servo, AC servo, stepper motor), by the transmission type (e.g., geared, direct-drive, harmonic-drive, timing-belt, chain and sprocket, tendoned, and traction-drive or friction-drive), by capacity and accuracy (e.g., heavy-duty industrial robots and microminiaturised finger robots), and by mobility.

Operation and Control

Robotic tasks can be grouped broadly into gross manipulation tasks and fine manipulation tasks. Control of the motion trajectory of the end effector of a robot is directly applicable to tasks in the first category. Examples of such tasks are seam tracking in arc welding, spray painting, contour cutting (e.g., laser and water jet) and joining (glueing, ultrasonic and laser merging), and contour inspection (e.g., ultrasonic, electromagnetic, and optical). Forcing and tactile considerations are generally crucial to tasks in the second category. Parts assembly, robotic surgery, machining, forging, and engraving are examples of fine manipulation tasks. It is intuitively clear that gross manipulation can be accomplished through motion control. But force control (including compliance control) also would be needed for accurate fine manipulation, particularly because small motion errors can result in excessive and damaging forces in this class of tasks.

For predefined gross-manipulation tasks, a robot is usually taught the desired trajectory either by using a mechanical input device such as a teaching pendant or joystick, or by off-line programming. Precise path planning and continuous path generation are essential in trajectory tracking applications. For tasks such as pick-and-place operations where the end positions (and orientations) are of primary interest, point-to-point interpolation may be employed. Trajectory segmentation and segmental interpolation also are commonly used in continuous trajectory control. Once an end-effector trajectory is specified, the desired joint trajectories may be determined by direct measurement using joint sensors during the teaching (learning) mode of operation, or alternatively by off-line computation using kinematic relations for the particular robot (see Chapter 2). During the task-repeat (playback) mode of operation, the desired joint trajectories are compared with the measured joint trajectories, and the associated joint error values are used by the manipulator servos for compensation.

Fine manipulation control which incorporates force and tactile information is generally more complex. Dexterity comes into play quite prominently here and conventional control techniques have to be augmented by more sphisticated control approaches such as hybrid force/position control, active compliance control and impedance control (Horn and Raibert, 1978; Mason and Salisbury, 1985; Whitney, 1987; de Silva, 1988).

The present generation of robotic manipulators, including many research robots, use hard control algorithms at the direct digital control level, with classical servo control as the norm (de Silva and Van Winssen, 1987). Since "intelligence" is frequently advocated as a requirement for future generations of industrial robots, there has been some research interest in incorporating heuristics and knowledge-based control into robot control systems. Knowledge and intelligence may be introduced directly into the controller of a robot, or may be distributed in components such as sensors (both internal and external), end effectors, and programmable fixtures. A major proportion of the research on intelligent control in robotics has gone into incorporating knowledge based control at the integrated system level, for example, in a flexible manufacturing cell, rather than at the robot level. Some work in incorporating knowledge based control into robot controllers, has been unfortunately devoted to adding a soft controller directly into the servo loop, without first exploring whether such a control scheme is appropriate.

Except in academic and research environments, a robot user is normally buffered from the intricate and complex programming which is needed to implement various control strategies. A typical user would program a robot through an appropriate high-level programming language, using simple English-like commands (Fu, et. al., 1987).

1.4 Robot Dynamics

Real-time control of a robotic manipulator, at typical bandwidths of the order of 100 Hz, is complicated due to the complex dynamics which are present in the manipulator. This complexity is due to many causes such as nonlinearity, dynamic coupling, high-order dynamics, and time variation of system parameters (de Silva, 1986). Nonlinearities are of several types including physical (e.g., coulomb friction and backlash), dynamic (e.g., centrifugal and coriolis accelerations), and geometric (e.g., trigonometric functions which enter into a robot model due to the link-to-link coordinate transformations which are needed for dynamic formulation). Dynamic coupling will result from geometric coupling as well as through such dynamic phenomena as coriolis accelerations. A high dynamic order of robotic manipulator may be the direct result of the total degrees of freedom; for instance, in an interactive operation of more than one robot, or due to the inclusion of modes of motion contributed by structural flexibility. Dynamic fixturing and pay load changes are major causes of fast (and large) parameter variations in a robot. A realistic dynamic model of a robotic manipulator has to include all these characteristics.

For a robotic task specified by an end-effector trajectory, a basic problem in manipulator control is to determine the torques for the revolute joints and the forces for the prismatic joints which will maintain the robot in the prescribed trajectory. This "inverse dynamics" problem is what must be addressed in this context. The formulation of manipulator dynamics is usually accomplished in two stages; first, a kinematic formulation, and next a dynamic (or kinetic) formulation (Asada and Slotine, 1986). This development is outlined in Chapter 2.

Inverse Kinematics

In the first stage of dynamic formulation, the end-effector trajectory is transformed (resolved) into joint trajectories. This is the "inverse kinematics" problem. Several cartesian coordinate frames are used for this purpose; frames fixed to bodies such as manipulator base (base frame), hand or end effector (end-effector frame), a link at each joint (joint frames), and even frames fixed at the centroids of manipulator links (link frames). A robotic task is normally specified with respect to a world frame which is an inertial frame fixed in the work space. Often, in non-mobile robots, the world frame is identical to the base frame. On defining the forward direction as that from the end effector to the base, a coordinate transformation can be established from a particular joint frame (j) to the one ahead of it (j-1). The Denavit-Hartenberg notation (1955) is used in practice to define the joint frames. Both frame position (origin) and frame orientation have to be expressed, and this information is usually given by a single (homogeneous) transformation matrix A_j (Paul, 1981). The associated joint coordinate q_j (relative translation for a prismatic joint or relative rotation for a revolute joint) is the variable quantity in a homogeneous transformation matrix. The product

$$I = A_1(q_1)\, A_2(q_2)\, ...A_n\,(q_n)$$ (1.1)

for an n degree-of-freedom robot gives the coordinate origin and the direction cosines of the three axes of the end effector frame with respect to (and expressed in) the base frame. There are six independent values (world coordinates) associated with this matrix for each point (both position and orientation) of the end-effector trajectory. Since these values are specified by the particular task (i.e., they are known quantities), the inverse kinematics problem is to determine the corresponding n joint coordinates $q_1, q_2, ..., q_n$. This problem will not have a unique solution in general (and is not trivial) because the associated relations are strongly nonlinear and coupled. Partial decoupling can be achieved by appropriate choices of robot geometry, for example, by the use of a spherical wrist. Nevertheless, general solutions are not possible, though customised solutions are available for specific manipulator geometries. The problem becomes more complicated in the presence of redundant kinematics. In this case, the manipulator has more than six degrees of freedom ($n > 6$), and generally an infinite number of solutions then will exist for the inverse kinematics problem. Optimisation constraints can be employed to obtain unique kinematic solutions with redundant manipulators (Asada and Slotine, 1986; de Silva, et. al., 1988). Kinematic redundancy provides additional dexterity to a robot and it can be effectively used, for instance, in obstacle avoidance (Yoshikawa, 1984).

Suppose that the end-effector trajectory of a robot is expressed using a sixth order column vector **r** containing three position elements and three orientation elements, expressed in terms of a suitable coordinate frame.

For an incremental joint motion $\delta \mathbf{q}$, the associated end-effector motion $\delta \mathbf{r}$ is given by

$$\delta \mathbf{r} = \underline{\mathbf{J}} \, \delta \mathbf{q} \qquad\qquad (1.2)$$

where $\underline{\mathbf{J}}$ is the manipulator jacobian. Equation (1.2) also expresses the end-effector velocity $\mathbf{v} = \dot{\mathbf{r}}$ in terms of the joint speeds $\dot{\mathbf{q}}$. Inverse dynamics require the solution of joint speeds for a specified end-effector velocity. Again, for a manipulator with kinematic redundancy, a unique solution does not exist in general, and additional constraints could be imposed. In the absence of kinematic redundancy, an additional problem arises at those geometric configurations where the jacobian $\underline{\mathbf{J}}$ becomes singular. Physically this singularity implies that the end-effector motion components are not totally independent; some end-effector motions are not feasible at singular configurations. Indeed, if kinematic redundancy is available, it can be utilised to avoid singular configurations, for a specified end-effector trajectory. Typically, however, singular configurations are avoided by a proper planning of task trajectories.

The relationship between joint accelerations $\ddot{\mathbf{q}}$ and end-effector acceleration $\mathbf{a}$ is obtained by differentiating equation (1.2); thus

$$\mathbf{a} = \underline{\mathbf{J}} \ddot{\mathbf{q}} + \dot{\underline{\mathbf{J}}} \dot{\mathbf{q}} \qquad\qquad (1.3)$$

It follows that the inverse of the manipulator jacobian has to be computed to determine joint speeds and joint accelerations for a specified end-effector trajectory. A major computational burden results, and so these computations should be performed off line whenever feasible. Solution of the inverse kinematics problem provides the $\mathbf{A}_j$ matrices and the joint motions $\dot{q}$ and $\ddot{q}$.

Inverse Dynamics (Recursive Formulation)

Once $\underline{A}_j$, $\dot{q}$, and $\ddot{q}$ are known, the second stage of the inverse dynamics formulation is reached. In this stage, the formulation can proceed in two ways; by using the Lagrangian formulation, or by using the Newton-Euler formulation.

In the Lagrangian formulation, the manipulator lagrangian is expressed as the difference between the kinetic energy and potential energy, and the generalised forces associated with the incremental generalised coordinates δq_j are expressed by Lagrangian equations of motion. These equations then provide the forces (or torques) which are needed to drive the manipulator. The conventional formulation (Uicker, 1965; Kahn, 1969) of Lagrangian dynamics contains three summations over 1 to n for each joint force (torque) term, and consequently $O(n^4)$ computations (additions and multiplications) are needed to compute all n drive force terms. This order can be reduced to $O(n)$ however, through an ingenious recursive formulation (Hollerbach, 1980). In this formulation, the dynamic equations are first expressed in terms of the product matrix

$$\underline{W}_j = \underline{A}_1 \underline{A}_2 \ldots \underline{A}_j \tag{1.4}$$

Then using the fact that

$$\underline{W}_j = \underline{W}_{j-1} \underline{A}_j \tag{1.5}$$

it is possible to compute the $\underline{W}_j$ matrices and their time derivatives $\dot{\underline{W}}_j$

and $\ddot{\underline{W}}_j$ starting from the manipulator base and ending at the end effector; a backward recursion. This is feasible because the base kinematics are completely known in general, and further $\underline{A}_j$, $\dot{q}_j$, and $\ddot{q}_j$ are also known from the inverse kinematics. Next a forward recursion is performed, starting from the end effector and ending at the base, to successively compute the joint forces (torques) f_j associated with q_j. This again is feasible since the end-effector forces are known from the task specification. Further gains in computational efficiency are possible if 3 x 3 transformation matrices are used instead of the 4 x 4 homogeneous transformation matrices as discussed in Chapter 2 (Hollerbach, 1980).

In the Newton-Euler formulation, force vectors at link joints are expressed in terms of rectilinear velocities at centroids and the angular velocities of the manipulator links. Here also the necessary computations can be reduced to O(n) by using a recursive formulation (Luh, et. al., 1980). Specifically, a backward recursion is performed for the link angular velocities ω_j, angular accelerations $\dot{\omega}_j$, link centroidal velocities v_{cj}, and link centroidal accelerations a_{cj}, with known values for $\dot{q}_j$ and $\ddot{q}_j$, and subsequently a forward recursion is performed to compute joint forces f_j and joint moments n_j.

Friction and Backlash

In modeling and computing the joint forces of a manipulator, bearing friction and gear friction are usually represented by an equivalent viscous friction model, and backlash is usually neglected completely. These are not realistic assumptions, except for a few special types of manipulator; for instance, backlash is negligible in direct-drive arms. An accurate computation of the inverse dynamics would necessitate more realistic models.

A realistic friction model for robotic manipulator joints is given in Figure 1.2 (Shibley, 1988). The coefficient of friction is defined such that its product with an equivalent joint reaction gives the frictional force (torque) in the direction of ($-\dot{q}_j$). The coefficient of friction is known to vary with relative speed as shown by a broken line in the figure and this relationship can be approximated by two straight-line segments as indicated. Then frictional terms can be included in the inverse dynamics by modifying the Newton-Euler recursive formulation using the following computational steps:

1. Compute joint velocities and joint forces/torques (including reactions) using Newton-Euler recursions and neglecting joint friction.

2. Obtain the corresponding coefficient of friction for each joint using the data in Figure 1.2 (say, by table look-up or by using a programmed analytical   relationship).

3. Compute the frictional force (torque) associated with q_j, and modify the drive force (torque) accordingly.

Strictly speaking, the reactions themselves (in Step 1) would change due to the presence of friction, and hence further cycles of computation would be needed until the values converge. But, in practice, a single cycle is known to provide accurate results.

Backlash is another effect that can significantly affect the drive forces and torques of a robotic manipulator. If the backlash frequency is sufficiently higher than the control bandwidth, then backlash may be

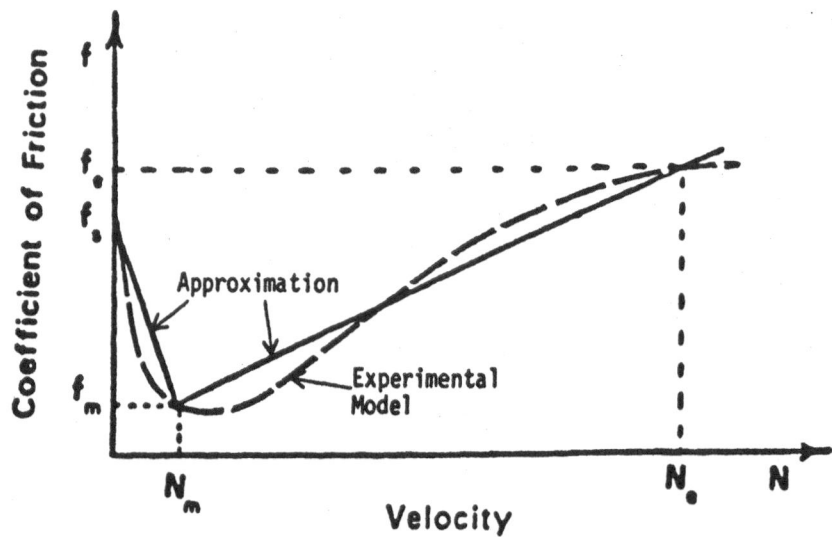

Figure 1.2 A Friction Model for a Joint of a Robot

treated as an unknown high-frequency disturbance. If on the other hand, the backlash frequency is low enough, it is proposed that the following steps be included in the Newton-Euler recursive formulation, to account for backlash:

1. Include gear stages in the dynamic formulation and compute the drive forces (torques) at every gear stage using the Newton-Euler recursive formulation.

2. If the drive torque at a gear stage changes sign, then there is backlash at that stage. Thus disengage that stage, and assume zero transmitted torque there.

3. Apply the Newton-Euler recursion to the last disengaged manipulator segment that includes the end effector, and compute the drive forces (torques) for the specified end-effector trajectory.

4. Compute motion of the remaining manipulator segments using drive forces (torques) computed in Step 1, and then use this information to check whether the segments will remain disengaged at the end of the current control cycle.

This proposed approach is presently under development.

1.5 Control of Robots

In an industrial environment it is quite unlikely that a robot will function independently purely as a stand-alone device. For example, as shown in Figure 1.3, a flexible manufacturing cell (FMC) intended for the production of small batches of various parts, might consist of one or more robots, several machine tools (milling machines, drills, forging machines, grinders, etc.), programmable fixtures (e.g., positioning tables, flexible

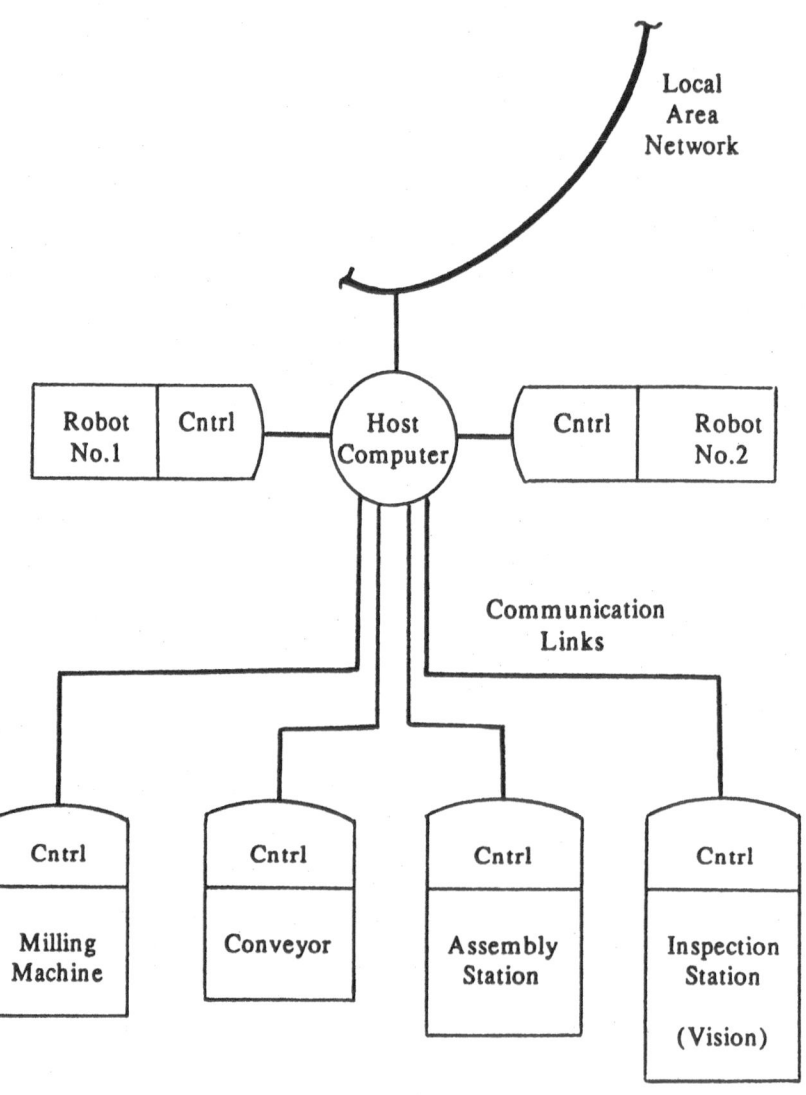

Figure 1.3 Schematic Diagram of a Flexible Manufacturing Cell

jigs), parts transfer mechanisms (e.g., conveyors, gantry mechanisms), and inspection and gauging stations (e.g., vision systems, laser-based gauging devices), all coordinated and managed by a cell host computer through direct communication links. Several cells linked via a local area network will form a flexible manufacturing system (FMS). A complex system of this nature is usually designed and operated as a hierarchical structure. A typical three-level hierarchy will employ a top-level supervisory computer to handle general managerial functions, task scheduling, coordination of machines and material flow, and fault management; it will employ an intermediate-level host computer to generate desired trajectories, tool speeds, feed rates and other reference signals for low-level controllers; and it will employ a set of low-level computers or hardware controllers for direct digital control. Each hierarchical level may also contain laterally distributed structures. Strictly, robot control considerations are not independent of the control of the overall hierarchical structure. However, control of a distributed and hierarchical system of this nature is beyond the scope of the present investigation. Our attention will be limited to the control of an individual robot, without regard to interactions of the robot with any other device or complex system.

Low-level control of an industrial robot is accomplished primarily through the servo control of individual joints in a conventional manner. This approach has several disadvantages (Horn and Raibert, 1978; Van Brussel and Vastmans, 1984; de Silva, 1984). Primarily, since servo parameters are set at constant values during each cycle of robot operation, these control parameters cannot adapt to compensate for robot nonlinearities and for parameter changes. Furthermore, an effective compensation for dynamic coupling among the joints of a manipulator is not possible through servo control.

A robot controller, particularly one for high speed operation, should possess several characteristics (de Silva, 1986): in particular, it should take into account any robot nonlinearities and dynamic coupling. In model-based control, this may be accomplished by using an appropriate dynamic model for the robot. An adequate controller should allow for complex tasks (trajectories), and for unexpected changes in task specifications during operation. It should also compensate for unknown factors such as model errors, external disturbances, and noise. It should be able to handle rapid parameter variations, including unplanned pay load changes and structural changes. It should cater for real-time control at high control bandwidths. This may be accomplished, for example, through the use of efficient algorithms (e.g., recursive relations and customised formulations), by employing off-line computation (e.g., gross drive force/torque computations), and perhaps by using special hardware implementations and parallel computation architectures.

Though not yet commercially implemented, many researchers have developed effective control algorithms which possess many of these desired qualities. Several such algorithms have been implemented on prototype research robots and tested. In computed-torque control, drive forces (torques) are computed using a suitable dynamic model and fed forward as a gross compensation signal. Some form of feedback control is essential with this scheme in order to avoid unstable behaviour due to unavoidable modeling errors. Computed-torque control with feedback servo control has produced quite satisfactory performance in experimental robots (Asada, et. al., 1983). With a customised algorithm implemented on an array processor, a force computation cycle for a six degree-of-freedom robot has been reported to take less than 2 ms (Khosla, 1986), providing a control bandwidth of the order of 250 Hz. Further improvements in computational speed can be realised by using table look-up procedures, but in this case at the expense of controller memory. In one such memory-intensive method, the position-dependent terms are precomputed and stored in the controller memory, with position coordinates as independent variables (Horn and Raibert, 1978). During control, at a given instant, the feedforward forces are computed through table look-up for that particular orientation of the manipulator in the configuration space. In addition to accounting for nonlinearities and dynamic coupling, the computed-torque control has the further advantage of control speed due to feedforward nature of the control.

In resolved-motion position control (Paul, 1977), a desired end-effector trajectory is transformed (resolved) into joint trajectories, then compared with measured joint responses, and servo loops are closed using the error signal. Velocity and acceleration loops may be added by measurement of the corresponding variables. In resolved-motion rate control (Whitney, 1969), first, measured joint trajectories are transformed and compared with the desired end-effector trajectory to determine the response error. Then the velocity necessary at the end effector to compensate for this error is determined, and transformed (resolved) back to joints. These computed joint velocities are then used to close a set of velocity loops. Acceleration loops may be also added by measurement. Resolved motion acceleration control is similar to resolved motion rate control, except that the end-effector acceleration needed to correct the trajectory error is estimated, and resolved back to the joints for closing acceleration loops there. In resolved-motion force control (Wu and Paul, 1982), the end-effector forces which are needed for a specific task are computed and resolved into joint forces (torques) using the transpose of the manipulator jacobian. This result is exact only in the static case. To compensate for any error due to dynamic effects, a force convergent control scheme is employed, requiring the measurement of end-effector forces and comparison with the computed forces. The algorithm does not provide a scheme to determine the controller parameter values of the force feedback loop. The need for end-effector force measurements further limits this approach.

Variable structure control or sliding mode control (Young, 1978; Slotine, 1985) is based on defining a time-varying surface such that trajectory following is equivalent to remaining on this sliding surface. A control law is derived from the specified requirement that the control action always drives the actual trajectory of the robot toward the sliding surface. This amounts to a switching controller, with the sliding surface functioning as the switching surface. The control law is based on a dynamic model for the robot, but high levels of modeling error are tolerated. Bounds for modeling error and external disturbances are used as parameters of the control law. In this sense, sliding mode control is quite robust. One drawback is the inherently chattering nature of the controller, resulting in high control activity and also possible excitation of high frequency dynamics that are neglected in the model. To reduce these problems a "boundary layer" is added to the sliding surface, and no switching is allowed if the response is retained within this layer. This improvement is brought about at the cost of a reduced control bandwidth. Unlike the computed-torque control, this method will retain some error even in the absence of modeling errors.

An adaptive control scheme which has been tested on a research manipulator of the industrial class is model-referenced adaptive control (Dubowski and Des Forges, 1979; Kornblugh, 1984). This approach stems from the fact that, in theory, if servo parameters can be continuously updated in real time, then idealised performance, as given by a suitable reference model driven by the same inputs as for the actual robot, may be realised. The error between the reference model response (desired output) and the robot response is used to drive an adaptive mechanism which updates the servo parameters (typically a proportional gain and a velocity feedback constant for each joint) such that a quadratic error index is minimised in the direction of the steepest slope of the performance index. Disadvantages of model-referenced adaptive control include the following: the feedback control structure has to be specified through some independent means; the adaptive law is dependent on the chosen reference model and is independent of robot model; control action has to be generated faster than the speed at which nonlinear terms in the robot change; and the adaptive law is derived by assuming some nonlinear terms of the robot to be constant.

Another adaptive control scheme which has been developed for robotic manipulators is the least squares adaptive control algorithm (de Silva, 1984; de Silva and Van Winssen, 1986). This is a model-based scheme which updates a feedback gain matrix so as to minimise a weighted quadratic error function at each point along the robot trajectory. The advantage of a computationally efficient recursive formulation of this control scheme is somewhat outweighed by the need to compute the state transition matrix of the robot about the desired trajectory, at each instant of gain update.

Several other control techniques including learning control (Freedy, 1973), time optimal control (Kahn and Roth, 1971; Lynch, 1981), and impedance control (Hogan, 1984) have been proposed in the past, for manipulator control. These techniques are considered to be primarily of pedagogical importance at the present time, and no commercial implementations are known. Hybrid force/position control where some degrees of freedom of a manipulator are force controlled and some others are motion controlled (Raibert and Craig, 1981) is particularly useful in fine manipulation. Nonlinear feedback control (Hemami and Camana, 1976; de Silva, 1988), and noninteracting control (Tokumaru and Iwai, 1971) bear special significance in the present work, because the underlying principle of decoupling control plays an important role in the control structure which is developed here. These considerations will be addressed in detail in Chapter 4.

1.6 Knowledge-Based Control

With steady advances in the field of artificial intelligence (AI), especially pertaining to the development of practical expert systems or knowledge systems, there has been considerable interest in using AI for controlling complex processes (Francis and Leitch, 1984; Goff, 1985; Isik and Meystel, 1986). The rationale for using intelligent control systems may be related to difficulties which are commonly experienced by practising control engineers: It is generally difficult to accurately model a complex process by a mathematical model or by a simple computer model. Even when the model itself is tractable, control of the process using a hard control algorithm might not provide satisfactory performance. Furthermore, it is commonly known that the performance of some industrial processes can be considerably improved through the control actions (tuning actions in particular) made by an experienced and skilled operator, and these actions normally cannot be formulated by hard control algorithms.

There are many reasons for the practical deficiencies of hard control algorithms. If the process model itself is inaccurate, model-based control can provide unsatisfactory results. Even with an accurate model, if the parameter values are partially known, ambiguous, or vague, then appropriate estimates have to be made. Algorithmic control which is based on such incomplete information will not usually give satisfactory results. The environment with which the process interacts may not be completely predictable, and it is normally not possible for a hard algorithm to accurately respond to a condition that it did not anticipate and that it could not "understand". A hard control algorithm will need a complete set of data to produce results, and indeed, program instructions and data are conventionally combined into the same memory of the controller. If some data were not available, say unexpectedly as a result of sensing problems, the inflexible algorithm will naturally fail.

Humans are flexible (or soft) systems. They can adapt to unfamiliar situations; they are able to gather information in an efficient manner and discard irrelevant details. The information which is gathered need not be complete or general and could be vague, for humans can reason, infer and deduce new information and knowledge. They have commonsense. They can make good decisions, and also can provide logical explanations for those decisions. They can learn, perceive, and improve their skills through experience. Humans can be creative, inventive, and innovative. It is a very challenging task to seek to develop systems which possess even a few of these "simple" human abilities. This is the challenge faced by workers in the field of AI and knowledge-based systems. However, humans have weaknesses too. For example, they tend to be slow, inaccurate, forgetful, and emotional. (In other areas of course, one might argue that these human qualities are strengths rather than weaknesses.)

In a knowledge-based system, one seeks to combine the advantages of a computer with some of the intelligent characteristics of a human for making inferences and decisions. It is clearly not advisable to attempt simply to mimic human thought process without carefully studying the needs of the problem at hand. In fact a faithful duplication is hardly feasible. Rather, in the initial stages of development of this field, it is adequate to develop special-purpose systems which cater for a limited class of problems. Knowledge-based control is one such development.

Knowledge Representation

According to Marvin Minsky of Massachusetts Institute of Technology, artficial intelligence is "the science of making machines do things that would require intelligence if done by men". At the basic level of implementation, however, what a machine (computer) does is quite procedural and cannot be considered intelligent. But in a wider perspective, and particularly viewing from outside and not from within the machine, any machine that appears to perform tasks in an intelligent manner can be considered as an intelligent machine. An appropriate representation of knowledge, including intuition and heuristic knowledge, is central to the development of machine intelligence and of knowledge-based systems.

In a knowledge-based system two types of knowledge are needed; knowledge of the problem (problem representation or modeling) and knowledge regarding methods for solving the problem. Ways of representing and processing knowledge include (Staugaard, 1987)

1. logic
2. semantic networks
3. frames
4. production systems.

In logic, the relevant knowledge is represented as a set of propositions. A proposition is a statement which is either true or false. In predicate calculus this idea is generalised by allowing for variables; then the truth value of a proposition will depend on the values assumed by the variables. Complex propositions can be formed using connectives such as AND, OR, NOT, EQUALS, and IMPLIES. Evaluating the truth values of predicates using known facts is the basic approach to problem solution using logic. One drawback of logic is its crisp nature and consequent inability to represent vague information which does not have sharp truth values. Fuzzy logic is a partial answer to this deficiency.

Semantic networks are used in the graphical representation of knowledge. Here knowledge objects are represented as nodes in a network, and their relationships are represented by arcs or lines linking the nodes. An arc represents a particular association between two objects. For example, a marriage can be represented by a "married to" arc joining a husband object and a wife object. Knowledge represented by a semantic network is processed using network searching procedures.

A frame is a data structure originally developed to represent expectational knowledge, or knowledge of what to expect when entering a given situation for the first time. Commonsense knowledge can be represented in this way and new information (new frames) can be interpreted using old information (old frames) in a hierarchical manner. A frame may contain context knowledge (facts) and action knowledge (cause-effect relations). Reasoning and problem solution using frames are done as follows: an external input (say, a sensor signal) triggers a frame according to some heuristics; then, slots in the frame are matched with context (current data including sensory inputs), which may lead to a possible updating of slot values in the current frame, and to triggering of subsequent frames once the conditions are matched.

Production systems or rule based systems are commonly used in knowledge representation and problem solution using artificial intelligence. The structure of a typical production system is shown in Figure 1.4. Here knowledge is represented by a set of rules (or productions) stored in the knowledge base. The data base contains the current data (or context) of the process. The inference engine is the control mechanism which coordinates and organises the sequence of steps used in solving a particular problem.

Two types of reasoning strategies are used in a production system:

1. forward chaining
2. backward chaining.

In forward chaining, the data base is searched to match an "if" (condition) part of a rule in the rule base, with known facts or data (context), and if a match is detected, that rule is fired (i.e., the "then" part or "action" part of the rule is activated). Actions could include creation, deletion, and updating of data in the data base. One action can lead to firing of one or more new rules. The inference engine is responsible for sequencing the matching (searching) and action cycles. In addition it is responsible for resolving conflicts which arise when more than one rule is matched. Methods of conflict resolution include firing the very first match, firing the matched rule that has the longest list of context elements (toughest match), firing the matched rule that has the longest high-priority elements (privileged rules), and firing the matched rule which has the most recent element in the context. Forward chaining is a bottom-up procedure. A production system that uses forward chaining is termed a forward

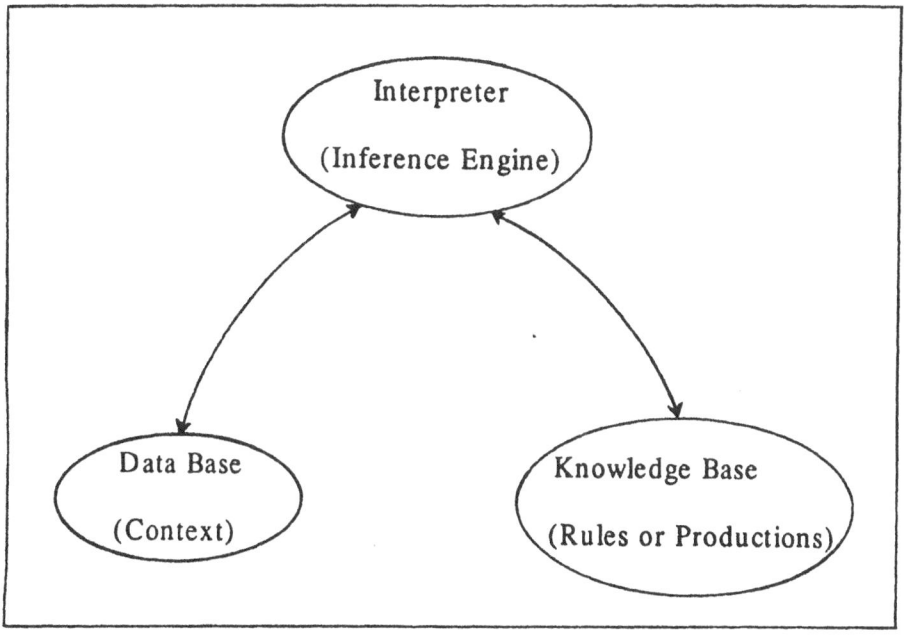

Figure 1.4 The Structure of a Production System

production system (FPS). This type of system is particularly useful in knowledge-based control.

In backward chaining, which is a top-down search process, a hypothesised conclusion is matched with rules in order to determine the context (facts) that supports the particular conclusion. If enough facts support a hypthesis, that hypothesis is accepted. Backward chaining is useful in situations which call for diagnosis, theorem proving, and generally in applications where a logical explanation has to be attached to each action. A production system which uses backward chaining is called a backward chaining system (BCS).

Fuzzy Knowledge

Formalisation of the concept of fuzzy sets and associated fuzzy reasoning (approximate reasoning) is due to Zadeh (1962). A fuzzy set does not have a crisp boundary. It can be represented by a membership function $\mu_A(x)$ which represents the grade of membership of the element x in the fuzzy set A. If $\mu_A(x) = 1$ for some value of x then this value is definitely a member (element) of A. Similarly, $\mu_A(x) = 0$ implies that the particular x is definitely outside A. A value within the range $0 < \mu_A(x) < 1$ means that the membership of x in A is vaguely defined. In this manner a vague or inexact quantity can be represented by a membership function with an associated fuzzy set. Note that such a membership function is a "possibility function" and not a probability function. It indicates the degree of possibility that a particular item is a member of the set A.

Another interpretation of fuzziness has been given (Sugeno, 1977). Consider a crisp (nonfuzzy) set A. Suppose that the membership of A is not known (but fixed). Then, one can define a grade of certainty $g_x(A)$ that a particular item x is a member of A. This is more like a probability that x is inside A, and so is different from Zadeh's original interpretation of fuzziness. Generality, ambiguity, uncertainty, and imprecision are terms which are often used in the context of fuzziness (see for instance Tong, 1977). But, strictly speaking in Zadeh's sense, these terms have meanings different from the formal definition of fuzziness. Generality can be associated with the use of a single symbol to represent more than one element. Ambiguity is attributed to the presence of more than one interpretation for a particular situation or quantity. Uncertainty is associated with probability. Precision is defined in terms of tolerance, and the tolerance limits involved are crisp and not fuzzy.

In the work presented here, Zadeh's interpretation of fuzziness is used and the theory involved is as presented by Dubois and Prade (1980), and is extensively used. The relevant concepts of fuzzy logic, particularly those useful in fuzzy control, are outlined in Chapter 3. The material presented in Chapter 3 is well-known (Mamdani and Gaines, 1981; Gupta and Sanchez, 1982; Zimmerman, et. al., 1984); however, new graphic representations, simplifying interpretations, and examples have been added in order to clarify the underlying concepts.

Humans freely use fuzzy representations in linguistic descriptions. For example, terms like tall, beautiful, more, and fast are naturally fuzzy. This vagueness in some linguistic terms is an important and useful characteristic of any natural language. It follows that fuzzy concepts are useful in knowledge representation in soft systems. In a fuzzy system, knowledge is represented by conventional "if-then" rules relating fuzzy variables, as well as through algebraic relations of fuzzy parameters and variables. The primary task in conventional fuzzy control is to determine a membership function for the control decision (or control action). The compositional rule of inference is utilised for this purpose. Since the quantities involved in a fuzzy reasoning procedure are generally fuzzy and the resulting inference (control decision) is also fuzzy, this procedure is often termed approximate reasoning. The related considerations are reviewed in Chapter 3.

Fuzzy Control

It is common knowledge in process control practice that when an experienced control engineer is included as an advisor to a control loop, significant improvements in the system performance are usually possible. Generally, control setting, tuning, and other control actions carried out by a human operator are implemented not through hard algorithms but rather qualitatively using linguistic rules which are based on knowledge and experience. For example, an expert might teach a process operator the necessary control actions using a set of protocols containing linguistic fuzzy terms such as "fast", "small", and "accurate". A practical difficulty invariably arises since except in very low bandwidth processes, human actions are not fast enough for this approach to be feasible. Furthermore, it is not economical to dedicate a human expert to every process since human experts are very hard to come by. In fuzzy control, linguistic descriptions of human expertise in controlling a process are represented as fuzzy rules or relations, and this knowledge base is used, in conjunction with some knowledge of the state of the process (say, of measured response variables), by an inference mechanism to determine control actions at a sufficiently fast rate.

Fuzzy control has been applied in a number of processes ranging from ship autopilots and pilot-scale steam engines to cement kilns and robotic manipulators (Van Amerongen, et. al., 1977; Mamdani, 1977; Holmblad and

Ostergaard, 1981; Scharf and Mandic, 1985; Hirota, et. al., 1985). One obvious drawback of such applications is that fuzzy control laws are implemented at the lowest level; within a servo or direct-digital-control (DDC) loop, generating control signals for process actuation directly through fuzzy inference. In high-bandwidth processes such as robotic manipulators, this form of fuzzy control implementation would require very fast and accurate control, in the presence of strong nonlinearities and dynamic coupling. Another drawback of this direct implementation of fuzzy control is that the control signals are derived from inferences which are fuzzy, thereby directly introducing errors into the control signals. A third argument against the conventional, low-level implementation of fuzzy control is that in a high-speed process, human experience is gained not through manual, on-line generation of control signals in response to process output, but typically through performing parameter adjustments and tuning (manual adaptive control) operations. Hence it can be argued that, in this case, the protocols for generating control signals are established not through direct experience but by some form of correlation. Of course, it is possible to manually examine input-output data recorded from a high-speed process. But, here, on-line human interaction is not involved and again the experience gained would be indirect and somewhat artificial. The work described here will take these shortcomings of conventional fuzzy control into account in developing an alternative control structure for robotic manipulator control.

The starting point of conventional fuzzy control is the development of a rule base using linguistic descriptions of control protocols, say, of a human expert. This step is analogous to the development of a hard control algorithm and the identification of parameter values for the algorithm, in a conventional control approach. The rule base consists of a set of **if-then** rules relating fuzzy quantities which represent process response (outputs) and control inputs. During the control action, process measurements (context) are matched with rules in the rule base, using the compositional rule of inference, to generate fuzzy control inferences. This inference procedure is clearly analogous to feedback control in a hard control scheme.

A rule in a fuzzy rule base is generally a fuzzy relation R_i of the form:

$$\text{If } A_i \text{ then if } B_i \text{ then } C_i \qquad (1.6)$$

where, A_i and B_i are fuzzy quantities representing process measurements (say, process error and change in error) and C_i is a fuzzy quantity representing a control signal (say, change in process input). Since these fuzzy sets are related through **if-then** implications and since an implication operation for two fuzzy sets can be interpreted as a "minimum operation" on the corresponding membership functions (see Chapter 3), the membership function of this fuzzy relation may be expressed as:

$$\mu_{Ri}(x,y,c) = \min(\mu_{Ai}(x), \mu_{Bi}(y), \mu_{Ci}(c)) \qquad (1.7)$$

The individual rules in the rule base are joined through **ELSE** connectives which are "unions" (**OR** connectives). Hence, the overall membership function for the complete rule base (relation R) is obtained using "supremum" operations on membership functions of the individual rules; thus

$$\mu_R (x,y,c) \ = \ \sup_i \ \mu_{Ri} (x,y,c) \qquad\qquad (1.8)$$

In this manner, membership function of the entire rule base can be identified using membership functions of the response variables and control inputs. Note that a fuzzy knowledge base is a multi-dimensional array (a 3-D array in the case of equation(1.8)) of membership function values. This array corresponds to a fuzzy control algorithm.

Once a fuzzy control algorithm of the form given by equation(1.8) is obtained, we need a procedure to infer control actions using process measurements, during control. Specifically, suppose that fuzzy process measurements A' and B' are available. The corresponding control inference C' is obtained using the *compositional rule of inference* (i.e., inference using the composition relation, as outlined in Chapter 3). The applicable relation is:

$$\mu_{C'} (c) \ = \ \sup_{x,y} \ \min (\mu_{A'} (x) , \mu_{B'} (y) , \mu_R (x,y,c)) \qquad (1.9)$$

Note that in fuzzy inference the fuzzy sets A' and B' are jointly matched with the fuzzy relation R. This corresponds to an **AND** operation and hence "min" operation applies for the membership functions. For a given value of control action c, the resulting fuzzy sets are then mapped from a three-dimensional space ($X \times Y \times Z$) on to a one-dimensional space (Z) of control actions. This mapping corresponds to a set of **OR** connectives and hence "sup" operation applies to the membership function values, as expressed in equation(1.9).

Actual process measurements are crisp. Hence they have to be fuzzified in order to apply the compositional rule of inference. This is done by representing a measurement by a singleton membership function having a unity value at the measurement element and zero values elsewhere. Conversely, the control action has to be a crisp value. Hence, each control inference C' has to be defuzzified so that it could be used to control the process. Several methods are available to accomplish this. In the mean of maxima method, the control element corresponding to the maximum grade of membership is used as the control action. If more than one element with maximum membership value is present, the mean of these values is used. In the centre of gravity method, elements of the *support set* of the control inference fuzzy set are weighted by their membership values and averaged. This weighted control action is known to provide a sluggish, yet more robust control. Schematic representation of a fuzzy controller of the type described here is shown in Figure 1.5. Here, application of the compositional rule of inference has been interpreted as a rule-matching procedure (also see Chapter 3).

There are several practical considerations of fuzzy control that have not been addressed in the above discussion. Since representation of the rule base R by an analytical function is unrealistic and infeasible in general, it is customary to assume that the fuzzy sets involved in R have discrete and finite universes (or, at least, discrete and finite support sets). As a result, process response measurements have to be quantised. Hence, at the outset, a decision has to be made as to the element resolution (quantisation error) of each universe. This resolution governs the cardinality of the sets and in turn the size of the multi-dimensional membership function array of a fuzzy rule base. It follows that computational effort, memory and storage requirements, and accuracy are directly affected by quantisation resolution. Since process measurements are crisp, one method to reduce real-time computational overhead is to

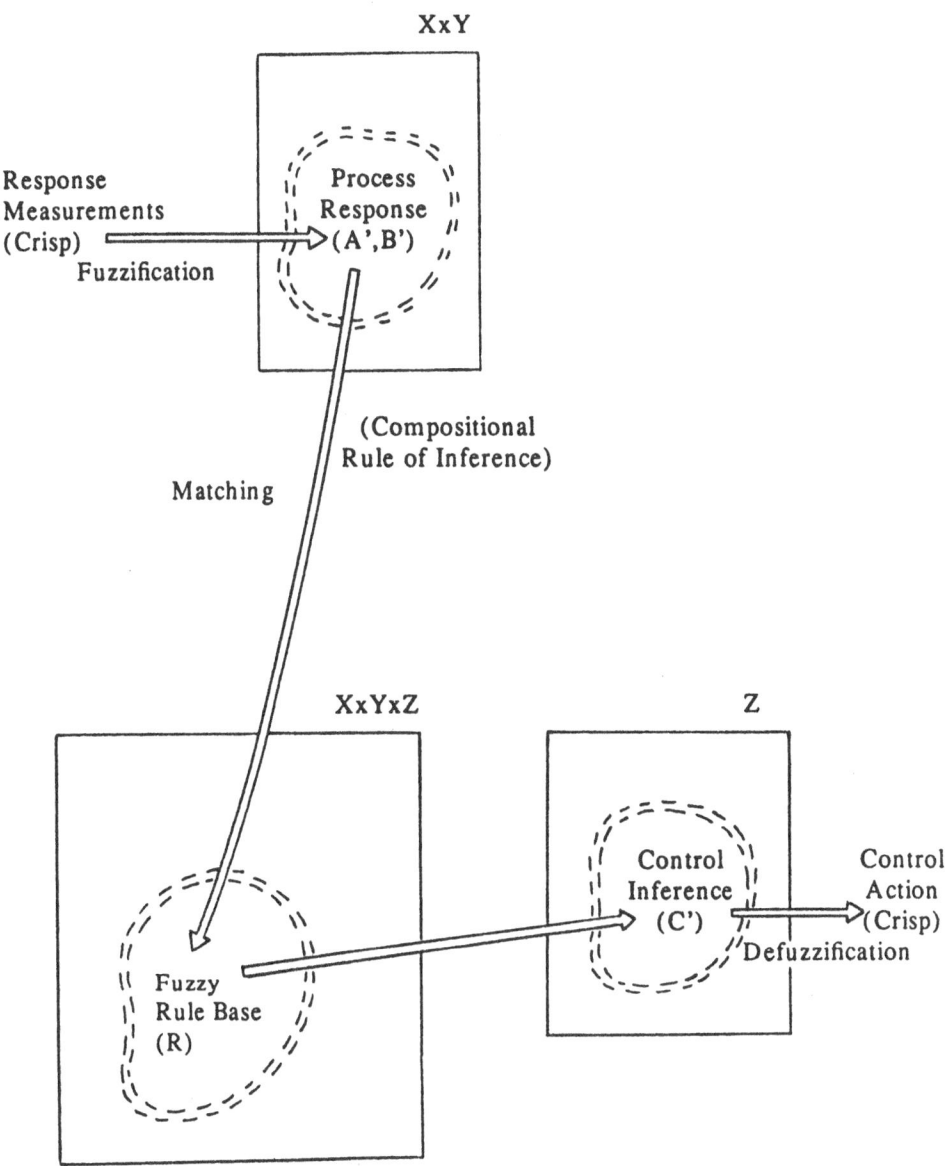

Figure 1.5 Schematic Representation of a Fuzzy Controller

precompute a decision table relating quantised measurements to crisp control actions. The main disadvantage of this approach is that it does not allow for modifications (e.g., rule changes and quantisation resolution adjustments) during operation. Another practical consideration is the selection of a proper sampling period, in view of the fact that process responses are generally analogue signals. Factors such as process characteristics, required control bandwidth, and the processing time needed for one control cycle have to be taken in to account in choosing a sampling period. Scaling or gain selection is another important consideration. For reasons of processing efficiency, it is customary to scale the process variables and control signals in a fuzzy control algorithm. Furthermore, adjustable gains can be cascaded with these system variables so that they may serve as tuning parameters for the controller. A proper tuning algorithm would be needed, however. A related consideration is real-time modification of a fuzzy rule base. Specifically, rules may be added, deleted, or modified on the basis of some self-organisation scheme (Procyk and Mamdani, 1979). For example, using a model for the process and making assumptions such as input-output monotonicity, it is possible during control, to trace the rules in the rule base that need attention. The control decision table can be modified accordingly.

The steps of fuzzy control, as described above, can be summarised in the following manner. A fuzzy control algorithm is developed first, according to the following four steps:

1. Develop a set of linguistic control rules (protocols).
2. Develop a set of discrete membership functions for process output variables and control input variables.

3. Using fuzzy implication on each rule in 1 and using 2, obtain the multi-dimensional array R_i of membership values for that rule.

4. Combine the relations R_i using fuzzy connectives (**AND, OR, NOT**) to obtain the overall fuzzy rule base (relation R).

Then, control action may be determined in real time as follows:

1. Fuzzify the measured process variables, as fuzzy singletons.

2. Match the fuzzy measurements obtained in 1 with the membership array of the fuzzy rule base (obtained in the previous Step 4), using the compositional rule of inference.

3. Defuzzify the control inference obtained in 2 (either the mean of maxima method or the centre of gravity method may be used here).

These steps reflect the formal procedure in fuzzy control. There are several variations. For example, a much faster approach would be to develop a crisp decision table by combining the four steps of fuzzy algorithm development and the first two steps of control, and using this table in a table look-up mode to determine a crisp control action during operation.

2. DYNAMIC FORMULATION OF ROBOT BEHAVIOUR

2.1 Introduction

This chapter deals with several kinematic and kinetic considerations which are important in the control of robotic manipulators. In kinematic modeling of robots, we are interested in expressing end effector motions in terms of joint motions. This is the direct problem in robot kinematics. The inverse-kinematics problem is concerned with expressing joint motions in terms of end-effector motions. This latter problem is in general more complex. In robot dynamics (kinetics), the direct problem is the formulation of a model as a set of differential equations for robot response, with joint forces/torques as inputs. Such models are useful in simulations and dynamic evaluations of robots. The inverse-dynamics problem is concerned with the computation of joint forces/torques using a suitable robot model, with the knowledge of joint motions. The inverse problem in robot dynamics is directly applicable to computed-torque control (also known as feedforward control), and also somewhat indirectly to the nonlinear feedback control method employed here.

2.2 Kinematic Formulation

Each degree of freedom of a robotic manipulator has an associated joint coordinate q_i. The robot is actuated by driving its joints. But a robotic task is normally specified in terms of end-effector motions. The end effector of a robot can be represented by a cartesian coordinate frame fixed to it (a body frame). This frame can be represented as a coordinate transformation with respect to some inertial frame (the world coordinate frame), typically a frame fixed to the stationary base of the robot (the base frame). The basic kinematics problem in modeling a robot is the expression of this coordinate transformation in terms of the joint coordinates q_i.

Homogeneous Transformation

Consider the cartesian frame (x_0, y_0, z_0) shown in Figure 2.1. If this frame is rotated about the z_0 axis through an angle θ_1, we get the cartesian frame (x_1, y_1, z_1) shown in the figure. The coordinate transformation associated with this frame rotation may be represented by the transformation matrix:

$$\underline{R}_1 = \begin{bmatrix} \cos\theta_1 & -\sin\theta_1 & 0 \\ \sin\theta_1 & \cos\theta_1 & 0 \\ 0 & 0 & 1 \end{bmatrix} \qquad (2.1)$$

We can make several important observations concerning $\underline{R}_1$. The columns of this matrix give the direction cosines of the axes of the new frame, expressed in the old frame. Consider any arbitrary vector r whose components are expressed in the new frame. Then, if we premultiply r by $\underline{R}_1$, we get the components of the same vector expressed in the old frame.

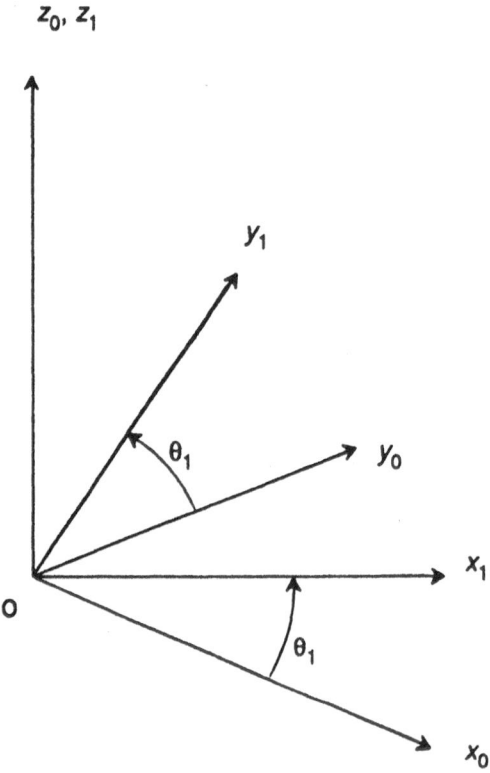

Figure 2.1   A Coordinate Transformation

Note that the same observations hold if the frame rotation were to be made about any arbitrary axis through the origin, not just z_0. In summary, we can make the following general statements:

1. A coordinate transformation $\underline{R}$ represents a rotation of a coordinate frame to a new position.

2. The columns of $\underline{R}$ give the direction cosines of the new frame axes expressed in the old frame.

3. Premultiplication of a vector **r** by $\underline{R}$ is equivalent to fixing **r** in the old frame and rotating the entire unit to the new frame position.

4. The product $\underline{R}$**r** gives the components of the vector **r** expressed in the old frame, **r** itself giving the componets of the vector in the new frame.

Suppose that the new frame (x_1, y_1, z_1) is next rotated to another position represented by the cartesian frame (x_2, y_2, z_2). The columns of the corresponding transformation matrix $\underline{R}_2$ give the direction cosines of the axes of (x_2, y_2, z_2) when expressed in the frame (x_1, y_1, z_1). It follows that the matrix product $\underline{R}_1\underline{R}_2$ gives the direction cosines of the axes of (x_2, y_2, z_2) expressed in the original frame (x_0, y_0, z_0). If the transformation $\underline{R}_2$ represents a rotation expressed in the frame (x_0, y_0, z_0), then the direction cosines of the final frame axes, expressed in the original frame (x_0, y_0, z_0), are given by the columns of the product $\underline{R}_2\underline{R}_1$. These ideas can be extended to a product of more than two transformation matrices.

The foregoing discussion considered rotations about an axis through the origin of a coordinate frame. Now let us consider pure translations. Consider a vector **r**, as in Figure 2.1, having three components expressed in a cartesian frame. Let us augment this vector with a unity element, to form the 4th order column vector r_a, given by:

$$r_a = \begin{bmatrix} r \\ 1 \end{bmatrix} \qquad (2.2)$$

Now consider a 4 x 4 matrix $\mathbf{I}$ given by:

$$\mathbf{I} = \begin{bmatrix} \mathbf{1} & \mathbf{p} \\ 0 & 1 \end{bmatrix} \qquad (2.3)$$

in which $\mathbf{1}$ denotes the 3 x 3 identity matrix and **p** is a vector expressed in the original cartesian frame, representing a pure translation. It is easy to verify that the product $\mathbf{I}r_a$ is given by:

$$\mathbf{I}r_a = \begin{bmatrix} r + p \\ 1 \end{bmatrix} \qquad (2.4)$$

It follows that the matrix $\mathbf{I}$ can be considered as a transformation matrix which represents a pure translation. Since this is a 4 x 4 matrix, if we are to combine rotations and translations into a single transformation, we must first convert the 3 x 3 rotation matrix R into an equivalent 4 x 4 matrix. Since vectors are augmented by a unity element in this approach, it is easily seen that the corresponding 4 x 4 rotation matrix is:

$$B_a = \begin{bmatrix} R & 0 \\ 0 & 1 \end{bmatrix} \qquad (2.5)$$

where 0 denotes a null column or row of compatible order. Now suppose that we translate a frame through vector p and then rotate the resulting frame about an axis through the origin of this new frame according to R. The overall transformation A is given by:

$$A = I B_a \qquad (2.6)$$

By direct matrix multiplication we get:

$$A = \begin{bmatrix} R & p \\ 0 & 1 \end{bmatrix} \qquad (2.7)$$

There is no rotation from the original frame to the intermediate frame. Hence, the direction cosines of the axes of the final frame, expressed in either the intermediate frame or the original frame, are given by the columns of R. It follows that the transformation matrix A contains all the information about the final frame, expressed in terms of the original frame. Specifically, p gives the *position* of the frame origin and R gives the *orientation* of the frame. Matrix A is a unified or "homogenised" representation of translations and rotations of a coordinate frame. For that reason $\underline{A}$ is known as a 4 x 4 *homogeneous transformation* matrix.

As a matter of interest, suppose that we first rotate the frame and then translate the resulting frame through **p** (of course, expressed in the intermediate frame with respect to which the translation is made). Then the overall homogeneous transformation matrix becomes:

$$
B_a I = \begin{bmatrix} \underline{R} & \underline{R}\mathbf{p} \\ 0 & 1 \end{bmatrix} \qquad (2.8)
$$

Indeed, this result is compatible with equation(2.7) because $\underline{R}\mathbf{p}$ is the translation expressed in the original coordinate frame.

The Denavit-Hartenberg Notation

To formulate robot kinematics, we wish to present a homogeneous transformation matrix representing a general coordinate transformation from one link of a robot to an adjacent link. For this purpose, body frames, fixed to links of the robot, are chosen according to the Denavit-Hartenberg notation. This is explained in Figure 2.2. Note that joint i joins link i-1 with link i. Frame i, which is the body frame of link i, has its z axis located at joint i+1. If the joint is *revolute*, then the joint rotation is about the z axis. If the joint is *prismatic*, the joint translation is along the z axis. It is seen from Figure 2.2 that frame i can be obtained by transforming frame i-1 as follows:

1. Rotate frame i through θ_i about the z axis.

2. Translate the new frame through d_i along the z axis.

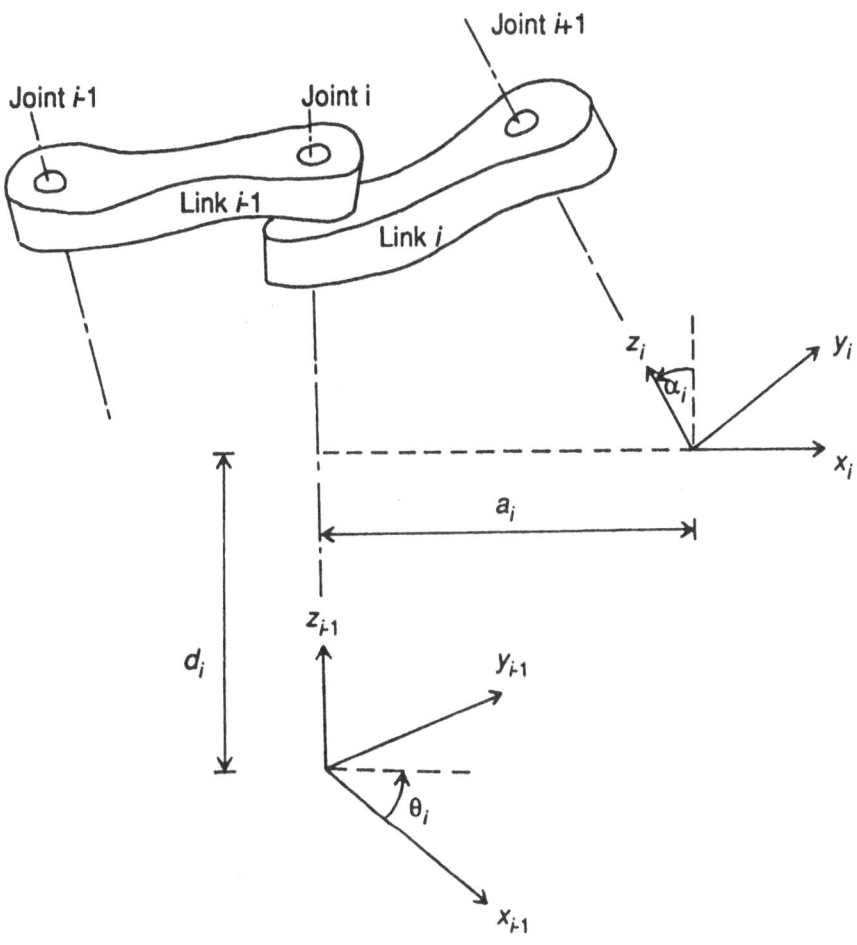

Figure 2.2   The Denavit-Hartenberg Notation

3. Translate the new frame through a_i along the new x axis.

4. Rotate the new frame through α_i about the current x axis.

Note that all these movements are carried out in the positive sense of a right-handed cartesian frame. The homogeneous transformation matrix $\underline{A}_i$ which represents the overall link-to-link transformation is obtained by the product of the four 4 x 4 transformation matrices corresponding to the above four steps, when taken in the proper order. It can be easily verified that this matrix is given by:

$$\underline{A}_i = \begin{bmatrix} \cos\theta_i & -\sin\theta_i\cos\alpha_i & \sin\theta_i\sin\alpha_i & a_i\cos\theta_i \\ \sin\theta_i & \cos\theta_i\cos\alpha_i & -\cos q_i\sin a_i & a_i\sin\theta_i \\ 0 & \sin\alpha_i & \cos\alpha_i & d_i \\ 0 & 0 & 0 & 1 \end{bmatrix} \qquad (2.9)$$

For a revolute joint, the joint coordinate would be:

$$q_i = \theta_i \qquad (2.10)$$

and, for a prismatic joint, the joint coordinate would be:

$$q_i = d_i \qquad (2.11)$$

with the remaining parameters in $\underline{A}_i$ kept constant. Hence the only variable in $\underline{A}_i$ is q_i.

The base frame, frame 0, is assumed fixed. This is taken as the inertial frame with respect to which a robotic task is specified. For a n degree-of-freedom robot, the body frame of the end effector is frame n, and this frame moves with the end effector. It follows that the position and orientation of the end-effector frame, expressed in the base frame, is given by the columns of the overall homogeneous transformation matrix $\underline{I}$;

$$\underline{I} = \underline{A}_1(q_1)\,\underline{A}_2(q_2) \cdots \underline{A}_n(q_n) \tag{2.12}$$

Equation(2.12) represents the kinematic formulation for a robotic manipulator.

Inverse Kinematics

Typically, a robotic task is specified in terms of the $\underline{I}$ matrix in equation(2.12). Since the drive variables are the joint variables, for the purposes of actuating and controlling a robot, it is necessary to solve equation(2.12) and determine the joint motion vector **q** corresponding to a specified $\underline{I}$. This is the inverse kinematics problem associated with a robot.

Since six coordinates are needed to specify a rigid body (or a body frame) in the three-dimensional space, $\underline{I}$ is specified using six independent quantities (typically, three position coordinates and three angles of rotation). It follows that equation(2.12) in general represents a set of six algebraic equations. These equations contain highly nonlinear trigonometric functions (of coordinate transformations) and are coupled.

Hence a simple and unique solution for q might not exist even in the absence of redundant kinematics ($n = 6$). Some simplification is possible by proper design of robot geometry. For example, by using a spherical wrist so that three of the six degrees of freedom are provided by three revolute joints whose axes coincide at the wrist of the end effector, it is possible to decouple the six equations in equation(2.12) into two sets of three simpler equations. In general, however, one must resort to numerical approaches to obtain the inverse-kinematics solution. In the presence of redundant kinematics ($n > 6$), an infinite set of solutions would be possible for the inverse-kinematics problem. In this case, it is necessary to employ a useful set of constraints for joint motions in order to obtain a unique solution.

Differential Kinematics

The jacobian matrix J of a robot is given by the relation

$$\delta r = J \, \delta q \qquad\qquad (2.13)$$

where, r is a sixth order vector, of which the first three elements represent the end-effector position (distance coordinates) and the remaining three elements represent the end-effector orientation (angles). As discussed in Chapter 1, it is important to determine J and its inverse in the computation of joint velocities and accelerations. This is the basic problem in differential kinematics, for a robotic manipulator.

Consider Figure 2.3 which uses the Denavit-Hartenberg notation. In

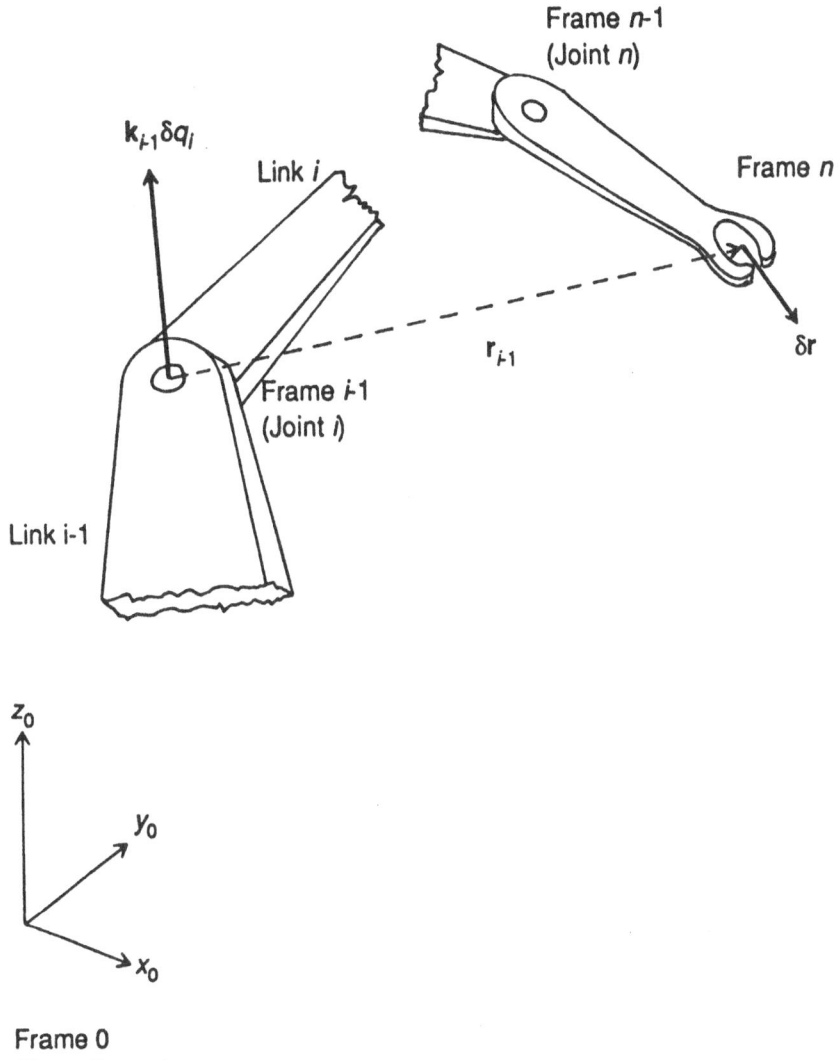

Figure 2.3   The Representation of Differential Kinematics

particular $k_{i\text{-}1}$ is a unit vector representing the axis of motion (rotation or translation) of joint i, expressed in the base frame. This is the z axis of the local frame (frame i -1). Vector $r_{i\text{-}1}$ is the position vector from joint i to the end-effector frame, expressed in the base frame. Also, δr is an incremental motion at the end effector, again expressed in the base frame, caused by an in incremental joint motion δq_i. It is easy to see that, if joint i is prismatic:

$$
\delta r \;=\; \begin{bmatrix} k_{i\text{-}1} \\ 0 \end{bmatrix} \delta q_i
\tag{2.14}
$$

and if the joint is revolute:

$$
\delta r \;=\; \begin{bmatrix} k_{i\text{-}1} \times r_{i\text{-}1} \\ k_{i\text{-}1} \end{bmatrix} \delta q_i
\tag{2.15}
$$

Now in view of equation(2.13), the right-hand-side vectors of equations (2.14) and (2.15) give the i th column of the jacobian matrix J, depending on whether the joint is prismatic or revolute. In this manner, the jacobian matrix can be constructed for a robotic manipulator.

If the manipulator does not contain redundant kinematics ($n = 6$), J would be a square matrix. Then the jacobian can be inverted, provided that it is not singular at the particular position of robot. In the case of a redundant manipulator ($n > 6$), however, additional constraints have to be introduced to joint motions in order to obtain a generalised inverse for J.

2.3 Inverse Dynamics

Robot dynamics can be formulated using either the Lagrangian method or the Newton-Euler method. The first approach is relatively more convenient to formulate and implement. But the tradeoff is that physical insight and part of the useful information (e.g., reaction forces at joints which are useful in computing friction and backlash, see Chapter 1) are lost in the process. An excellent description of the second approach is found in (Asada and Slotine, 1986). The following formulation is restricted to the Lagrangian approach.

In the Lagrangian approach to the inverse-dynamics problem of a robot, first, kinetic energy T and the potential energy U are expressed in terms of joint motion variables q_i and $\dot{q}_i$. Next, the lagrangian

$$L = T - U \tag{2.16}$$

is formed and the Lagrange equations of motion are written according to

$$\frac{d}{dt} \frac{\delta L}{\delta \dot{q}_i} - \frac{\delta L}{\delta q_i} = f_i \tag{2.17}$$

$$i = 1, 2, ..., n$$

where f_i are the input forces/torques at the joints; the generalised forces in the Lagrange formulation. By adopting this approach, we can obtain the following set of equations for f_i :

$$f_i = \sum_{j=i}^{n} \left[\sum_{k=1}^{j} \left(\mathrm{tr}\left(\frac{\delta W_j}{\delta q_i} \, \mathbb{J}_j \, \frac{\delta W_j}{\delta q_k}^{\mathsf{T}} \right) \ddot{q}_k \right. \right. +$$

$$\left. \sum_{p=1}^{j} \mathrm{tr}\left(\frac{\delta W_j}{\delta q_i} \, \mathbb{J}_j \, \frac{\delta^2 W_j}{\delta q_k \delta q_p}^{\mathsf{T}} \right) \dot{q}_k \dot{q}_p \right) - m_j \, g^{\mathsf{T}} \frac{\delta W_j}{\delta q_i} \, r_j \left. \right]$$

$$i = 1, 2, ..., n \qquad (2.18)$$

where, W_j is the homogeneous transformation which gives the position and orientation of frame j, when expressed in the base frame; thus

$$W_j = A_1(q_1)A_2(q_2) \cdots A_j(q_j) \qquad (2.19)$$

$\mathbb{J}_j$ is the moment of inertia matrix of link j expressed in the body frame j of the link, m_j is the mass of the link, r_j is the position vector of the centroid of link j expressed in and relative to frame j, and g is the gravity vector expressed in the base frame.

Note that equation(2.18) represents a set of nonlinear and coupled differential equations which can be put into the form:

$$M(q)\ddot{q} = n(q,\dot{q}) + f(t) \qquad (2.20)$$

where M is the inertia matrix of the robot and f is the vector of drive forces or torques. The vector n represents the nonlinear terms contributed by *coriolis* and *centrifugal* accelerations and gravity (Also see equation

(4.1) in Chapter 4). Nonlinearities are present in the terms involving both q and $\dot{q}$. Nonlinearities in q are caused by the coordinate transformations which are used in the dynamic formulation, which are *trigonometric* nonlinearities. These nonlinearities appear in the potential energy (gravity) term as well as in the kinetic energy (inertia) terms. Nonlinearities in $\dot{q}$ are quadratic functions which arise from centrifugal and coriolis acceleration components. Also, note that each computation of f_j involves three summations over a range of up to n, and that there are n such computations. It follows that the direct computation of the joint force vector f represents an $O(n^4)$ computation. This high order in joint force/torque computation is not acceptable in real-time control situations. A more efficient algorithm is needed. A recursive algorithm has been developed in (Hollerbach, 1980), which has reduced the order of computation to $O(n)$.

Recursive Formulation

Equation(2.18) may be written as

$$f_i = \sum_{j=i}^{n} \left[tr\left(\frac{\delta W_j}{\delta q_i} J_j \ddot{W}_j^T \right) - m_j g^T \frac{\delta W_j}{\delta q_i} r_j \right] \qquad (2.21)$$

$$i = 1, 2, ..., n$$

Suppose that we have computed A_j and its first and second derivatives with respect to q_j. Then f_i can be computed using the following backward recursion:

$$\underline{W}_j = \underline{W}_{j-1} A_j$$

$$\dot{\underline{W}}_j = \dot{\underline{W}}_{j-1} A_j + \underline{W}_{j-1} \frac{dA_j}{dq_j} \dot{q}_j$$

$$\ddot{\underline{W}}_j = \ddot{\underline{W}}_{j-1} A_j + 2\dot{\underline{W}}_{j-1} \frac{dA_j}{dq_j} \dot{q}_j + \underline{W}_{j-1} \frac{d^2 A_j}{dq_j^2} \dot{q}_j^2 +$$

$$\underline{W}_{j-1} \frac{dA_j}{dq_j} \ddot{q}_j$$

$$j = 2, 3, ..., n$$
$$(2.22)$$

starting with the initial condition $\underline{W}_1 = A_1$, and next performing the following forward recursion:

$$\underline{D}_i = \underline{J}_i \ddot{\underline{W}}_i^T + A_{i+1} \underline{D}_{i+1}$$

$$c_i = m_i r_i + A_{i+1} c_{i+1}$$
$$i = n-1, n-2, ..., 1 \qquad (2.23)$$

using the initial conditions $\underline{D}_n = \underline{J}_n \ddot{\underline{W}}_j^T$ and $c_n = m_n r_n$ and finally computing

$$f_i = tr\left(\frac{\delta \underline{W}_i}{\delta q_i} \underline{D}_i\right) - g^T \frac{\delta \underline{W}_i}{\delta q_i} c_i \qquad (2.24)$$

$$i = 1, 2, ..., n$$

Note that the derivatives $\delta \underline{W}_i / \delta q_i$ are known, because A_j and their derivatives are known.

3. FUZZY LOGIC

3.1 Introduction

Fuzzy logic is the logic which deals with fuzzy sets. The concepts of fuzzy sets and fuzzy logic are used in fuzzy control. The rule of fuzzy inference, namely, the *compositional rule of inference,* is particularly applicable in the context of fuzzy control. In this chapter, the mathematics of fuzzy sets is summarised, as this is necessary to understand the theory of fuzzy control. What is presented is an interpretation of some theoretical considerations described by Dubois and Prade (1980). Geometrical illustrations and examples have been added here, in order to facilitate the interpretation.

3.2 Fuzzy Sets

A fuzzy set is a set that does not have a sharp (*crisp*) boundary. In other words, there is a vagueness associated with the membership of elements in a fuzzy set. Consider a universe of discourse X whose elements are denoted by x. A fuzzy set A in X may be represented by a Venn diagram as in Figure 3.1(a). Generally, the elements x are not numerical quantities. But for analytical convenience the elements x are assigned real numerical values.

(a)

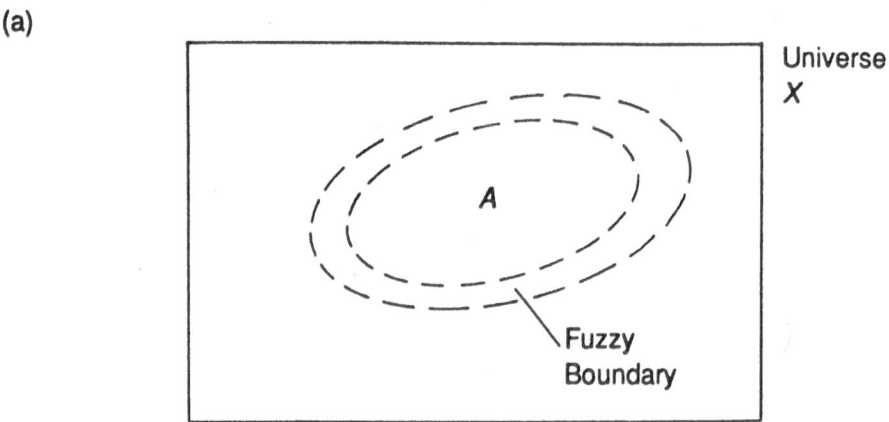

(b)

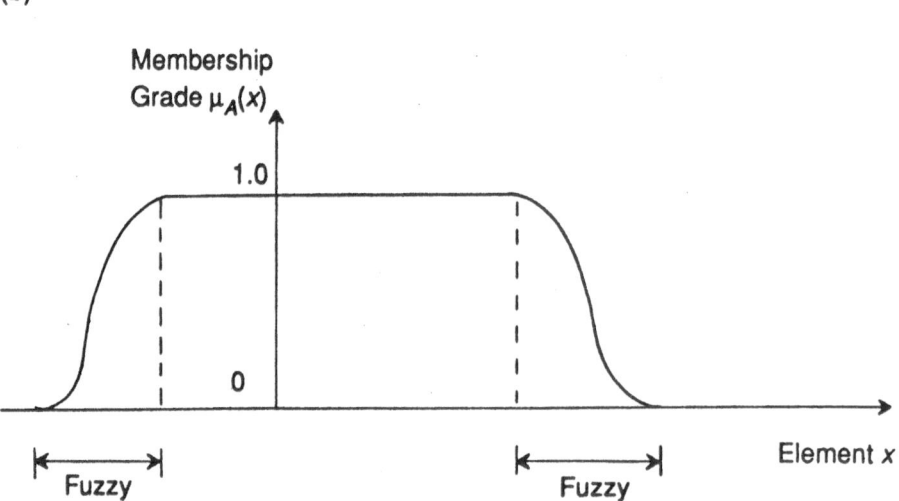

Figure 3.1 A Fuzzy Set
(a) Venn Diagram
(b) Membership Function

Membership Function

A fuzzy set may be represented by a membership function. This function gives the *grade (degree) of membership* within the set, of any elment of the universe of discourse. The membership function maps the elements of the universe on to numerical values in the interval [0, 1]. Specifically,

$$\mu_A (x)$$
$$X \longrightarrow [0,1] \tag{3.1}$$

where $\mu_A (x)$ is the membership function of the fuzzy set *A*. Note that a membership function is a possibility function and not a probability function. A membership function value of zero implies that the corresponding element is definitely not an element of the fuzzy set. A membership function value of unity means that the corresponding element is definitely an element of the fuzzy set. A grade of membership greater than 0 and less than 1 corresponds to a vague (fuzzy) membership, and the corresponding elements fall on the fuzzy boundary of the set. A typical membership function is shown in Figure 3.1(b).

Symbolic Representation

A universe of discourse, and a membership function which spans the universe completely defines a fuzzy set. We are not usually interested in the elements with zero grade of membership. The crisp set formed by the elements having a nonzero membership grade is termed the *support set* of the particular fuzzy set. The support set is a crisp subset of the universe.

Furthermore, a fuzzy set is clearly a subset of its support set.

A fuzzy set may be symbolically represented as:

$$A = \{ x : \mu_A(x) \} \tag{3.2}$$

If the universe is discrete with elements x_i then a fuzzy set may be specified using a convenient form of notation due to Zadeh, in which each element is paired with its grade of membership in the form of a "formal series" as:

$$A = \mu_A(x_1)/x_1 + \mu_A(x_2)/x_2 + \ldots + \mu_A(x_i)/x_i + \ldots$$

or

$$A = \sum_{x_i \in X} \frac{\mu_A(x_i)}{x_i} \tag{3.3}$$

If the universe is continuous, an equivalent form of notation is given in terms of a symbolic integration:

$$A = \int_{x \in X} \frac{\mu_A(x)}{x} \tag{3.4}$$

It is important to emphasise that both the series in (3.3) and the integral in (3.4) are symbolic shorthand forms of notation; to highlight this, no differential symbol d(.) is used in (3.4).

Examples

Three examples are now given to illustrate various types of fuzzy sets and their representations.

Example 1: Suppose that the universe of discourse X is the set of positive integers (or natural numbers). Consider the fuzzy set A in this discrete universe, given by the Zadeh notation:

$$A \; = \; 0.2/3 \; + \; 0.3/4 \; + \; 1.0/5 \; + \; 0.2/6 \; + \; 0.1/7$$

This set may be interpreted as a fuzzy representation of the integer 5.

Example 2: Consider the continuous universe of discourse X representing the set of real numbers. The membership function:

$$\mu_A(x) \; = \; 1/[1 + (x - a)^{10}]$$

defines a fuzzy set A whose elements x vaguely represent those satisfying the crisp relation $x = a$. This fuzzy set corresponds to a *fuzzy relation*.

Example 3: Linguistic terms such as "tall men", "beautiful women", "fast cars", and "slight increase" define fuzzy sets since their membership is subjective and vague.

3.3 Logical Operations

It is well known that the "complement", "union", and "intersection" of crisp sets correspond to the logical operations NOT, OR, and AND respectively, in the corresponding crisp proportional logic. Furthermore, the union of a set with the complement of a second set represents an "implication" of the first set by the second set. Set inclusion (subset) is a special case of implication in which the two sets belong to the same universe. These logical operations (connectives) have to be extended to fuzzy sets for use in fuzzy reasoning and fuzzy control. In fuzzy logic, these connectives have to be expressed in terms of the membership functions of the sets which are operated on.

Complement (NOT) A'

Consider a fuzzy set A in a universe X. Its complement A' is a fuzzy set whose membership function is given by:

$$\mu_{A'}(x) = 1 - \mu_A(x) \qquad\qquad \forall\, x \, \varepsilon \, X \qquad (3.5)$$

The complement corresponds to a NOT operation in fuzzy logic.

Union (OR) $A \vee B$

Consider two fuzzy sets A and B in the same universe X. Their union is a fuzzy set $A \vee B$. Its membership function is given by:

$$\mu_{A \vee B}(x) = \max(\mu_A(x), \mu_B(x)) \qquad\qquad \forall\, x \, \varepsilon \, X \qquad (3.6)$$

The union corresponds to a logical OR operation. The rationale for the use of "max" is that since the element x may be in one set or the other, the larger of the two membership grades should apply.

Intersection (AND) $A \wedge B$

Again consider two fuzzy sets A and B in the same universe X. Their intersection is a fuzzy set $A \wedge B$. Its membership function is given by:

$$\mu_{A \wedge B}(x) \ = \ \min\left(\mu_A(x), \mu_B(x)\right) \qquad \forall \ x \ \varepsilon \ X \qquad (3.7)$$

The intersection corresponds to a logical AND operation. The rationale for the use of "min" is that since the element x has to belong to both sets simultaneously, the smaller of the two membership grades should apply.

Implication (IF - THEN) $A \longrightarrow B$

Consider a fuzzy set A in a universe X and a second fuzzy set B in another universe Y. The fuzzy implication $A \longrightarrow B$ is a fuzzy relation in the cartesian product space $X \times Y$. There are several interpretations for this relation. Two commonly used relations for obtaining the membership function of the fuzzy implication are given below.

Method 1:

$$\mu_{A \longrightarrow B}(x, y) \ = \ \min\left(\mu_A(x), \mu_B(y)\right) \qquad (3.8)$$

$$\forall \ x \ \varepsilon \ X, \ \forall \ y \ \varepsilon \ Y$$

Method 2:

$$\mu_{A \to B}(x, y) = \min(1, (1 - \mu_A(x)) + \mu_B(y)) \qquad (3.9)$$

$$\forall x \in X, \ \forall y \in Y$$

Note that the first method gives an expression which is symmetric with respect to A and B. This is not intuitively satisfying because "implication" is not a commutative operation. In practice, however, this method provides good results. The second method has an intuitive appeal because in crisp logic, $A \to B$ has the same truth table as ((NOT A) OR B) and hence is equivalent. Note that, in equation(3.9), the membership function is upper-bounded to 1 using the *bounded sum* operation, as needed by definition. The first method is more commonly used because it is simpler and often provides more accurate results than the second method.

3.4 Fuzzy Relations

Consider two universes $X_1 = \{x_1\}$ and $X_2 = \{x_2\}$. A crisp set R consisting of a subset of ordered pairs (x_1, x_2) is a crisp relation in the cartesian product space $X_1 \times X_2$. An example is shown in Figure 3.2(a). Analogously, a fuzzy set R consisting of a subset of ordered pairs (x_1, x_2) is a fuzzy relation in the cartesian product space $X_1 \times X_2$. The relation R will be represented by the membership function $\mu_R(x_1, x_2)$. An example of a fuzzy relation is shown in Figure 3.2(b). This concept can be extended in a straightforward manner to fuzzy relations in the n-dimensional cartesian space $X_1 \times X_2 \times ... \times X_n$.

(a)

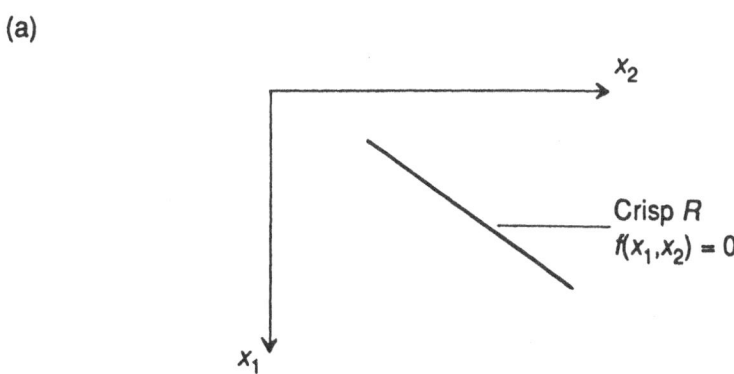

Crisp R
$f(x_1, x_2) = 0$

(b)

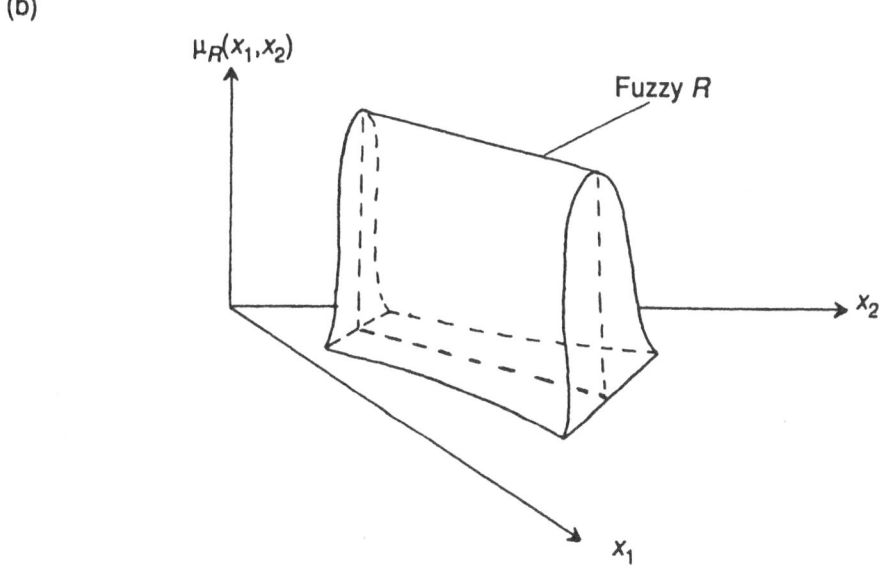

$\mu_R(x_1, x_2)$

Fuzzy R

x_2

x_1

Figure 3.2   (a) A Crisp Relation R in a Two-Dimensional Space (Plane)
(b) A Fuzzy Relation R in a Two-Dimensional Space

Example: Consider the fuzzy set R in the universe $X_1 \times X_2$, given by the membership function:

$$\mu_R (x_1, x_2) \; = \; 1/[1 + 100(x_1 - 3x_2)^4]$$

This is a fuzzy relation vaguely representing the crisp relation $x_1 = 3x_2$. In particular, note that all elements satisfying $x_1 = 3x_2$ have unity grade of membership and hence they are definitely in the set R. Elements satisfying, say, $x_1 = 3.1x_2$ have membership grades less than 1; their membership in R is vague. The further away the elements are from the straight line $x_1 = 3.1 \, x_2$, the more vague the membership of those elements in R.

Cartesian Product of Fuzzy Sets

Consider a crisp set A_1 in the universe X_1 and a second crisp set A_2 in another universe X_2. The cartesian product $A_1 \times A_2$ is a subset of the cartesian product space $X_1 \times X_2$, defined in the usual manner as shown in Figure 3.3(a). Next consider a fuzzy set A_1 in the universe X_1 and a second fuzzy set A_2 in another universe X_2. The cartesian product $A_1 \times A_2$ is then a fuzzy subset of the cartesian product space $X_1 \times X_2$. Its membership function is given by:

$$\mu_{A_1 \times A_2}(x_1, x_2) \; = \; \min (\mu_{A_1}(x_1), \; \mu_{A_2}(x_2)) \qquad (3.10)$$

$$\forall x_1 \; \varepsilon \; X_1, \; \forall x_2 \; \varepsilon \; X_2$$

(a)

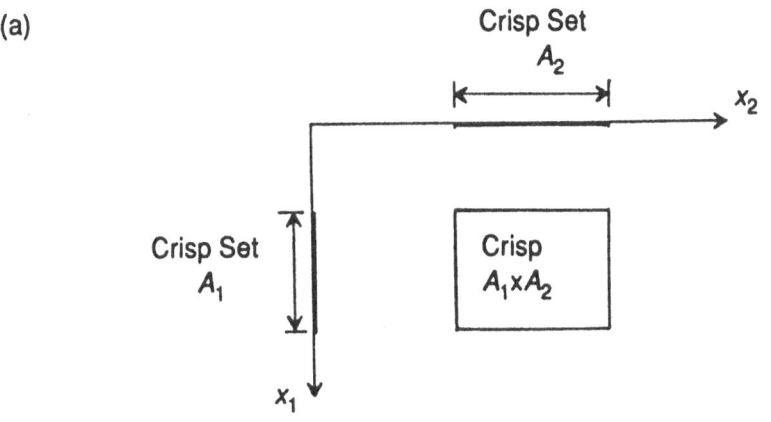

(b)

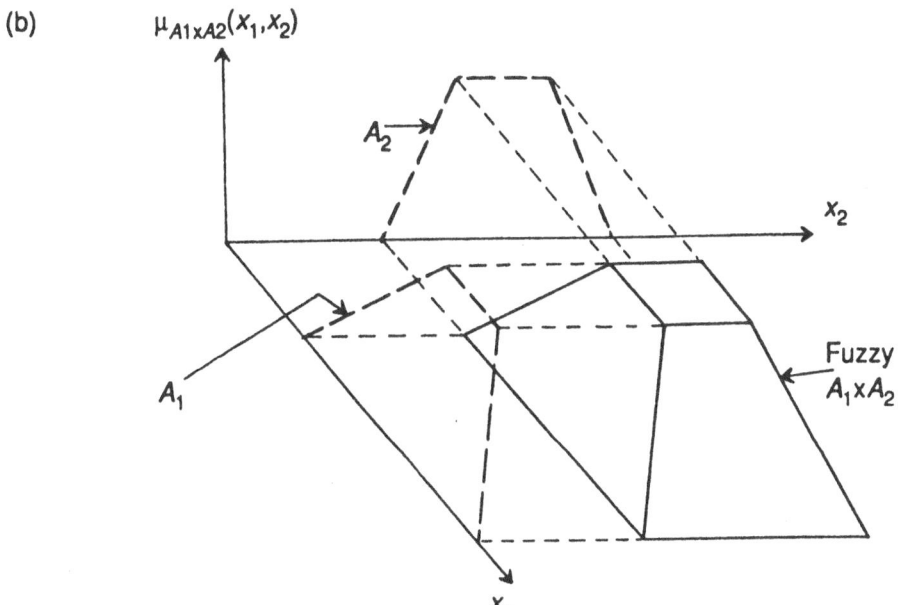

Figure 3.3 Cartesian Product $A_1 \times A_2$ (Relation) of
(a) Two Crisp Sets
(b) Two Fuzzy Sets

Note that the "min" combination applies here because each element (x_1, x_2) in the cartesian product is formed by taking both elements x_1 and x_2, not just one or the other. An example of a cartesian product of two fuzzy sets is shown in Figure 3.3(b).

The cartesian product of two fuzzy sets is a fuzzy relation, and is identical to the first method of fuzzy implication given by equation(3.8). The concept of cartesian product can be directly extended to more than two fuzzy sets.

Extension Principle

The extension principle was introduced by Zadeh to give a method for extending standard (non-fuzzy) mathematical concepts to their fuzzy counterparts. Consider the relation:

$$y = f(x_1, x_2, ..., x_r) \qquad (3.11)$$

where x_i are defined in the universes X_i, $i = 1, 2, ... , r$, and y is defined in the universe Y. This relation, generally, is a many-to-one mapping from the r-dimensional cartesian product space $X_1 \times X_2 \times ... \times X_r$ to the one-dimensional space Y:

$$X_1 \times X_2 \times ... \times X_r \longrightarrow Y \qquad (3.12)$$
$$(f)$$

Suppose that the elements x_i are restricted to crisp subsets A_i of X_i. Then the relation f maps elements $(x_1, x_2, ..., x_r)$ within the cartesian product $A_1 \times A_2 \times ... \times A_r$ onto a crisp set B which is a subset of Y. An example is given in Figure 3.4(a). Note that this is a many-to-one mapping. For instance, the entire quarter circle with centre (1,1) and radius 0.5 in the product space $A_1 \times A_2$ is mapped on to the single point 0.25 on the Y line.

This idea can be extended to the case where the sets A_i are fuzzy. The extension principle provides the means of doing this. Note that f is still a crisp relation. According to the extension principle, the fuzzy set B to which the elements y belong has a membership function given by:

$$\mu_B(y) = \sup_{\substack{(x_1, x_2 ..., x_r) \\ y = f(x_1, x_2, ..., x_r)}} \left\{ \min \left(\mu_{A_1}(x_1), \mu_{A_2}(x_2), ..., \mu_{A_r}(x_r) \right) \right\} \quad (3.13))$$

Note that the "min" operation applies first because the relation among A_i is the cartesian product (see equation(3.10). The "supremum" is applied over the mapping on to B, because more than one combination of $(x_1, x_2, ..., x_r)$ in the fuzzy space $A_1 \times A_2 \times ..., A_r$ will be mapped to the same element y in the fuzzy set B and the most possible mapping is the one with the highest membership grade. An example of a mapping from a two-dimensional fuzzy space $A_1 \times A_2$ on to a one-dimensional fuzzy set B is shown in Figure 3.4(b).

70

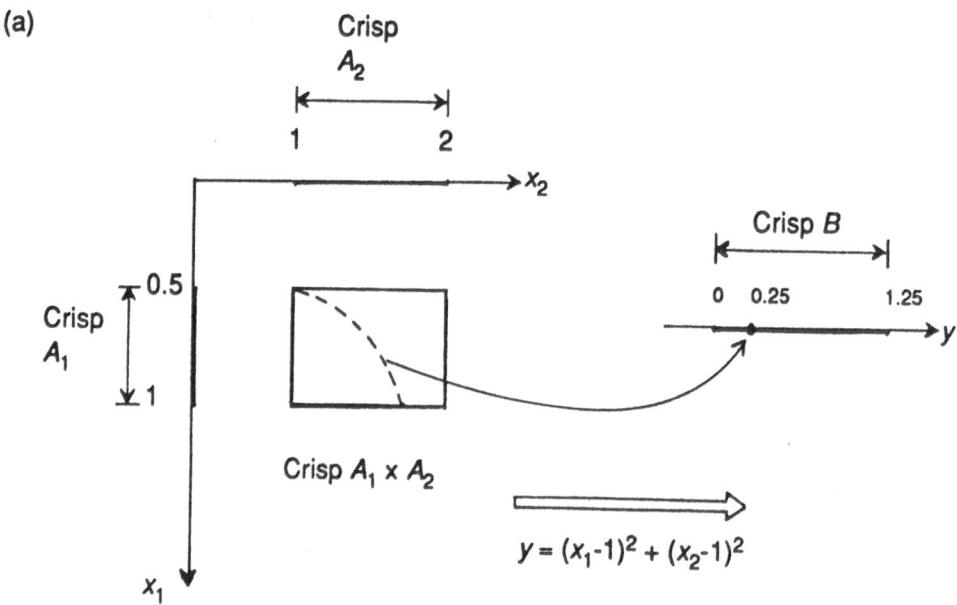

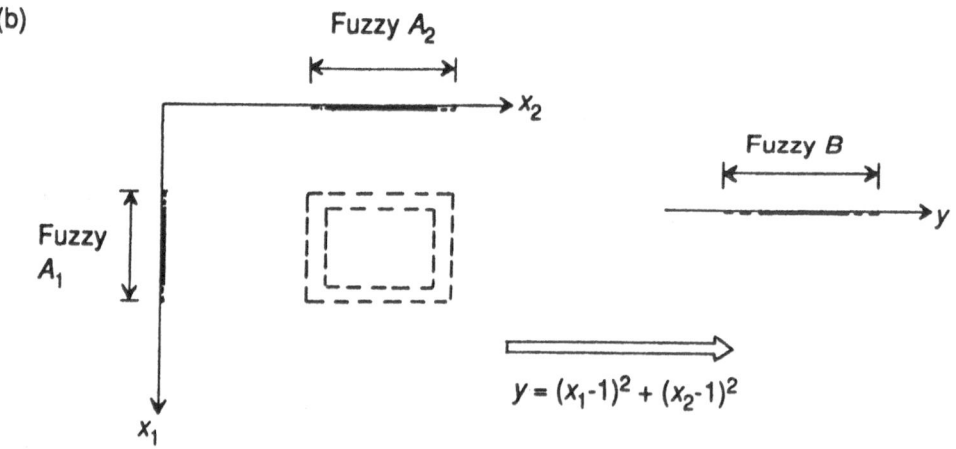

Figure 3.4 A Mapping from a Product Space to a Line
(a) An Example of Crisp Sets
(b) An Example of Fuzzy Sets
(Extension Principle)

3.5 Fuzzy Dynamic Systems

Consider a nonlinear and time-invariant dynamic system expressed in the discrete state-space form:

$$x_{n+1} = f(x_n, u_n) \tag{3.14}$$

$$y_n = g(x_n) \tag{3.15}$$

For the purpose of explaining the underlying concepts, only the scalar case is considered. The state variable x and the output variable y are assumed fuzzy, and the input variable u is assumed crisp. Generally in a fuzzy system, the state transition relation f and the output relation g will both be fuzzy relations. Suppose that the fuzzy sets corresponding to x_n, x_{n+1}, and y_n are $X(n)$, $X(n+1)$, and $Y(n)$ respectively. A typical simulation objective for a fuzzy dynamic system would be to determine the membership functions of $X(n+1)$ and $Y(n)$ once the membership functions of $X(n)$, f, and g are known. The extension principle is applicable here, allowing for the fact that the relations f and g are also fuzzy. Specifically, we have the following results:

$$\mu_{X(n+1)}(x_{n+1}) = \sup_{x_n \, \varepsilon \, X} \min \{ (\mu_{X(n)}(x_n), \mu_f(x_{n+1}, x_n, u_n)) \} \tag{3.16}$$

$$\mu_{Y(n)}(y_n) = \sup_{x_n \, \varepsilon \, X} \{ \min (\mu_{X(n)}(x_n), \mu_g(x_n, y_n)) \} \tag{3.17}$$

Note that X denotes the universe in which the state variable x lies; in this case the state space. It will be clear from the next section that "composition" operation is used in the above two results. Specifically, the fuzzy set $X(n+1)$ is obtained through *composition* of the fuzzy set $X(n)$ and the fuzzy relation f. Similarly, the fuzzy set $Y(n)$ is obtained through composition of the fuzzy set $X(n)$ and the fuzzy relation g. In equation(3.16) for example, "min" applies first because element x_n and relation f are matched through an AND operation. Next, "sup" applies because many elements x_n might be mapped to the same element x_{n+1} and hence the most desirable (possible) mapping has to be chosen. This is exactly the idea of composition.

3.6 Composition and Inference

Approximate reasoning is used in fuzzy inference and control. In particular, the *compositional rule of inference* is utilised. We have already introduced the concept of *fuzzy implication*. We shall start the present section by introducing the terms *projection*, *cylindrical extension*, and *join*, which will lead to the concept of *composition*. Finally the compositional rule of inference will be discussed, incorporating all these ideas.

Projection

Consider a fuzzy relation R in the cartesian product space $X_1 \times X_2 \times ..., \times X_n$. Suppose that the n indices are arranged as follows:

$$\{1, 2, ..., n\} \quad \longrightarrow \quad \{ i_1, i_2, ..., i_r, j_1, j_2, ..., j_m \} \quad (3.18)$$

Note that $r + m = n$ and that i and j denote the newly ordered set of n

indices. The projection of R on the subspace $X_{i1} \times X_{i2} \times ..., \times X_{ir}$ is denoted by:

$$\text{Proj } [R ; X_{i1} , X_{i2} ,...., X_{ir}]$$

This is a fuzzy set P and its membership function is given by:

$$\mu_P(x_{i1} , x_{i2} ,...., x_{ir}) = \sup_{x_{j1}, x_{j2},...,x_{jm}} \{ \mu_R(x_1 , x_2 , ... , x_n) \} \qquad (3.19)$$

The rationale for using the "supremum" operation on the membership function of R should be clear in view of the fact that we have a many-to-one mapping from a n-dimensional space to a r-dimensional space, with $r < n$.

As an example, consider the fuzzy set R shown in Figure 3.2(b) or the cartesian product $A_1 \times A_2$ of two fuzzy sets, shown in Figure 3.3(b), both of which are fuzzy relations in the two-dimensional space $X_1 \times X_2$. For instance, the projection of R on X_1 has a membership function which is exactly the projection of $\mu_R(x_1, x_2)$ on the μ-x_1 plane.

Cylindrical Extension

Consider the cartesian product space $X_1 \times X_2 \times ..., \times X_n$ and, suppose that the n indices are arranged as follows:

$$\{1, 2, ..., n \} \longrightarrow \{ i_1, i_2, ..., i_r , j_1, j_2, ..., j_m \}$$

Again note that $r + m = n$ and that i and j denote the newly ordered set of n indices. Now consider a fuzzy relation R in the subspace $X_{j1} \times X_{j2} \times ... \times X_{jr}$. Its cylindrical extension, denoted by $C(R)$, is given by

$$C(R) \quad = \quad \int_{X_1 \times X_2 \times ... \times X_n} \frac{\mu_R(x_{j1}, x_{j2}, ..., x_{jr})}{x_1, x_2, ..., x_n} \qquad (3.20)$$

Note that a cylindrical extension is a fuzzy set in the n-dimensional space and is the converse of projection. An example is given in Figure 3.5. Here a fuzzy set R in the universe X_1 has been cylindrically extended to a fuzzy set in the cartesian space $X_1 \times X_2$.

Join

Consider a fuzzy relation R in the subspace $X_1 \times X_2 \times ... \times X_r$ and a second fuzzy relation S in the subspace $X_m \times X_{m+1} \times ... \times X_n$ such that $m < r+2$. Note that the union of these two subspaces gives the space $X_1 \times X_2 \times ... \times X_n$. The join of the fuzzy sets R and S is a fuzzy set in the $X_1 \times X_2 \times ... \times X_n$. space, and is given by the intersection of their cylindrical extensions; thus:

$$Join(R,S) \quad = \quad C(R) \wedge C(S) \qquad (3.21)$$

$$in \ X_1 \times X_2 \times ... \times X_n$$

(a)

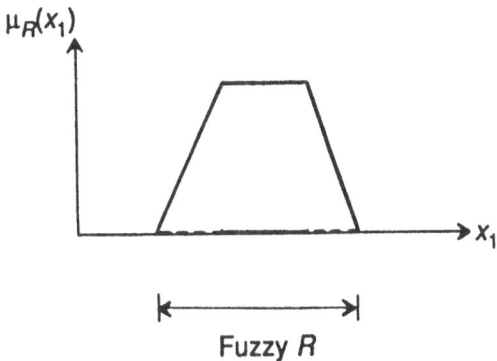

Fuzzy R

(b)

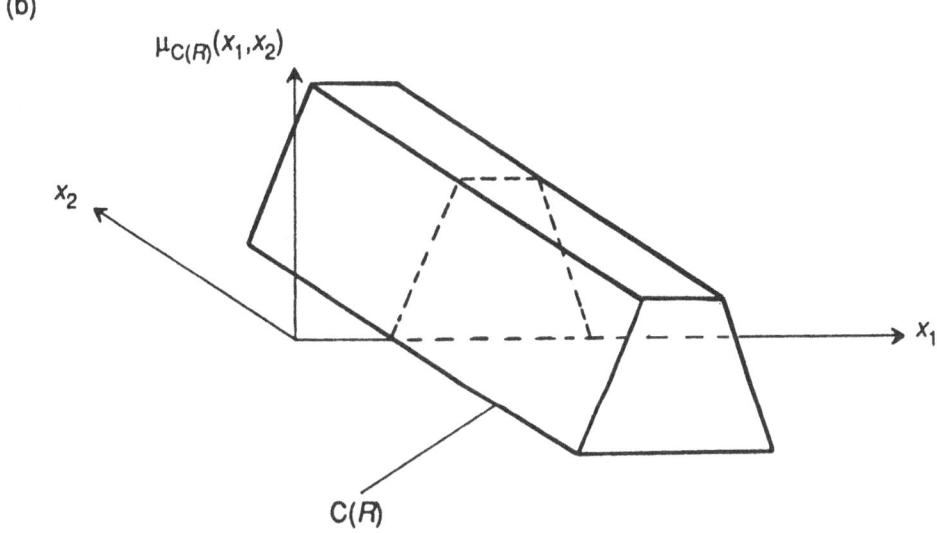

C(R)

Figure 3.5 The Cylindrical Extension
(a) A Fuzzy Relation (Set)
(b) Its Cylindrical Extension

Its membership function is given by:

$$\mu_{Join}(x_1, x_2, ..., x_n) = min (\mu_{C(R)}(x_1, x_2, ..., x_n) , \mu_{C(S)}(x_1, x_2, ..., x_n))$$

$$(3.22)$$

Note that "min" applies here because the intersection of two fuzzy sets is considered.

Composition

Consider a fuzzy relation (fuzzy set) R in the subspace X_1 x X_2 x ... x X_r and a second fuzzy relation (fuzzy set) S in the subspace X_m x X_{m+1} x ... x X_n such that $m < r +1$. Note that, unlike the previous case of *Join*, the two subspaces are never disjoint and hence their intersection is never null. But, as before, the union of the two subspaces gives X_1 x X_2 x ... x X_n. The composition of R and S is denoted by $R \circ S$ and is given by:

$$S \circ R = Proj [Join(R,S) ; X_1, ..., X_{m-1}, X_{r+1}, ..., X_n] \qquad (3.23)$$

Here we take the Join of the two sets, as given by equation(3.21) and then project the resulting fuzzy set on to the subspace formed by the disjoint parts of the two subspaces in which the fuzzy sets R and S are defined. The membership function of the resulting fuzzy set is obtained from the membership functions of R and S, while noting that "min" applies for Join and "supremum" applies for projection. Specifically,

$$\mu_{R \, oS} = \sup_{X_m, \, \dots, \, X_r} \{ \min (\mu_R , \mu_S) \} \qquad (3.24)$$

Composition can be interpreted as a matching of two fuzzy sets. Specifically, the two sets are combined first (Join) and then matched over their common subspace (supremum) giving a fuzzy set defined over the disjoint portions of the two subspaces (projection). This process of matching is quite analogous to matching data with the condition part of a rule in rule based control. In this regard, composition plays a crucial role in fuzzy inference and control. Note that composition is a commutative operation; $R \, o \, S = S \, o \, R$.

Example: As a simple example, consider a fuzzy relation R defined in the $X \times Y$ space and another fuzzy relation S defined in the $Y \times Z$ space. The composition of these two fuzzy sets is given by:

$$\mu_{R \, oS}(x,z) = \sup_{y \, \varepsilon \, Y} \{ \min (\mu_R(x,y) , \mu_S(y,z)) \} \qquad (3.25)$$

Compositional Rule of Inference

In knowledge based control systems, control knowledge is often expressed as rules of the form

" **If** output Y_1 is y_1 **then if** output Y_2 is y_2 **then** control C is c "

In fuzzy control, rules of this type are linguistic statements of expert knowledge in which y_1, y_2, and c are fuzzy quantities (e.g., small negative, fast, large positive). These rules are fuzzy relations that employ the fuzzy implication (**If-then**). The collective set of rules forms the knowledge base in fuzzy control. Let us denote the fuzzy relation formed by this collection of rules as the fuzzy set R. In fuzzy control, rules in R are first matched with available data (context). Next, a matched rule is fired thereby providing a control action. Usually, the context would be output measurements of the process and these are crisp quantities, and the control action which drives the process is a crisp quantity as well. But for a general consideration suppose that the data (context) are denoted by a fuzzy set D and the control action is denoted by a fuzzy set C. The compositional rule of inference states that:

$$C = D \circ R \qquad\qquad (3.26)$$

By using equation(A2.26) we can determine the membership function of the control action, using the knowledge of the membership functions of data and rule base. Specifically we have:

$$\mu_C = \sup_Y \ \min(\mu_D, \mu_R) \qquad\qquad (3.27)$$

This result follows directly from equation(3.24). Note that Y denotes the space in which the data D is defined, and it is a subspace of the space in which the rule base R is defined. Furthermore, since R consists of fuzzy implications, its membership function can be formed by the constituent membership functions using the "min" operation (see equation(3.8)). This method of obtaining R is analogous to model identification in conventional hard control.

Example: Suppose that a fuzzy set A represents the output of a process and that it belongs to a discrete and finite universe Y of cardinality (number of elements) 5. Furthermore, suppose that a fuzzy set C represents the control input to the process and that it belongs to a discrete and finite universe Z of cardinality 4. It is given that:

$$A = 0.2/y_2 + 1.0/y_3 + 0.8/y_4 + 0.1/y_5$$
$$C = 0.1/c_1 + 0.7/c_2 + 1.0/c_3 + 0.4/c_4$$

A fuzzy relation R is defined by the fuzzy implication $A \longrightarrow C$. The membership function of R is obtained using equation(3.8); thus:

$$\mu_R(y_i, c_j) = \begin{bmatrix} 0 & 0 & 0 & 0 \\ 0.1 & 0.2 & 0.2 & 0.2 \\ 0.1 & 0.7 & 1.0 & 0.4 \\ 0.1 & 0.7 & 0.8 & 0.4 \\ 0.1 & 0.1 & 0.1 & 0.1 \end{bmatrix}$$

This matrix represents the rule base in fuzzy control. Specifically, the information carried by the fuzzy rules $A \rightarrow C$ has been reduced to a relation R in the form of a matrix. Note that R is not a decision table for fuzzy control, even though a decision table can be derived using R by successively applying the compositional rule of inference to R for expected process responses.

Now suppose that a process measurement y_o is made and that the measurement is closest to the element y_4 in Y. This is a crisp set and it may be represented by a fuzzy singleton A_o with a membership function:

$$\mu_{Ao}(y_i) \; = \; [\, 0, 0, 0, 1\, , 0\,]$$

The membership function of the corresponding control inference C_o is obtained using the compositional rule of inference (equation(3.27)). It may be easily verified that:

$$\mu_{Co}(c_i) \; = \; [\, 0.1, 0.7, 0.8, 0.4\,]$$

This is simply the fourth row of μ_R. This fuzzy inference has to be defuzzified for use in process control, say, using the centre of gravity method.

3.7 Membership Function Estimation

Establishment of the membership functions of input and output variables is an important first step of fuzzy control. The simplest and most common way of determining a membership function is to first decide on a discrete universe of discourse for the particular fuzzy set and then guess a grade of membership for each value in this universe. A triangular function peaked at the most representative value of the fuzzy quantity is commonly used for this purpose. More formal and systematic methods are available for estimating membership functions of fuzzy quantities. Five such methods are outlined in this section.

Averaged Guess Method

In this method, several experts are asked to independently guess a membership function for the fuzzy quantity, and the data obtained in this manner are then averaged.

Distance Function Method

In this method, first a distance function $d(x)$ is estimated using common sense or available information. This function is a measure of the distance of an element x from the fuzzy set A. If a value of x is known to be definitely an element of A then the distance is taken to be zero. If a value is definitely not an element of A then its distance is taken as some upper-bound value (sup d). The membership function is given by the relation:

$$\mu_A (x) \;=\; 1 - \frac{d(x)}{\text{sup } d} \qquad\qquad (3.28)$$

An Intuitive Relation

It is intuitively clear that the rate of change of a membership function with respect to element value should increase with both the strength of belief and the strength of disbelief that the value belongs to the set. Analytically we can express this by:

$$\frac{d\mu_A(x)}{dx} = k\,\mu_A(x)\,(1 - \mu_A(x)) \qquad (3.29)$$

Direct integration of this relation gives:

$$\mu_A(x) = \frac{1}{1 + \exp(a - bx)} \qquad (3.30)$$

The parameters a and b have to be determined using additional knowledge of the problem.

Use of Binary Polling

In this method each member of a group of experts is asked the question whether a specified value x is an element of the fuzzy set, allowing only yes or no answers. This polling is repeated for all the values of the (discrete) universe of discourse. Then, the membership function is estimated as:

$$\mu_A(x) = \frac{\text{Number of yes answers}}{\text{Total number of answers}} \qquad (3.31)$$

Relative Preference Method

Consider a fuzzy set

$$A = \{ x_i : \mu_A(x_i) \} \qquad (3.32)$$

in the discrete and finite universe X of cardinality n. The relative preference of x_i over x_j for membership of A is denoted by p_{ij}. A reasonable measure for this quantity is given by

$$p_{ij} \;=\; w_i \,/\, w_j \qquad\qquad\qquad (3.33)$$

in which

$$w_i \;=\; \mu_A(x_i) \qquad\qquad\qquad (3.34)$$

The matrix $\underline{P}$ whose elements are given by equation(3.33), is said to be consistent. Note that $p_{ii} = 1$ and that $p_{ji} = 1/p_{ij}$. This matrix has the following properties:

1. All eigenvalues of $\underline{P}$ are zero except one which is n.
2. The eigenvector corresponding to this maximum eigenvalue (n) is the membership function vector **w** having the discrete grades of membership w_i as elements.

It follows that the membership function of a fuzzy set can be determined by first estimating a relative preference matrix $\underline{P}$ and then computing the eigenvector corresponding to its maximum eigenvalue. A measure of consistency of this estimate would be $\lambda_{max}(\underline{P})/n$ where $\lambda_{max}(\underline{P})$ is the maximum eigenvalue of $\underline{P}$. If this value is close to unity, the estimate is considered accurate.

4. CONTROL STRUCTURE

4.1 Introduction

Strong nonlinearities, dynamic coupling, high-order dynamics, and unknown influences were mentioned in Chapter 1, as reasons for control difficulties associated with robotic manipulators. Several hard algorithms intended for handling these difficulties were outlined and their shortcomings pointed out. Knowledge-based control, in particular fuzzy control, was discussed as a potentially powerful approach which can capture human experience and expertise in controlling complex processes, thereby circumventing many of the shortcomings of hard algorithmic control.

Most of the research effort in process control has been concentrated on hard algorithms, while in the realm of knowledge-based control, attention has been limited primarily to an exclusive use of soft techniques. In robot control research in particular, a few researchers have used soft techniques at the servo level (Scharf and Mandic, 1985; Hirota, et. al., 1985). Specifically, conventional fuzzy control, as outlined in Chapter 1, has been implemented directly in the servo loops. Results often show (Scharf and Mandic, 1985) that direct servo control outperforms sophisticated fuzzy control when implemented in this manner.

In robot control, the reason should be intuitively clear why knowledge-based controllers often do not give satisfactory results. For a robot, assuming that speed of operation is no obstacle, suppose that a human expert is called upon to generate control signals for the joint actuators. It is quite likely that an expert will resort to a hard algorithm to accomplish this, because it would be virtually impossible to generate appropriate drive signals in a heuristic way. On the other hand, if the expert interacts with the control loop in an advisory capacity, perhaps for tuning the control parameters, then generally the performance should improve. This possibility is supported by experience in the process control industry. This intuitive reasoning provides a rationale for the control structure for robot control which is proposed here.

4.2 The Control Hierarchy

The control structure now proposed is an attempt to combine the advantages of hard algorithmic control and knowledge-based soft control, for the control of robotic manipulators. In particular, hard algorithms are restricted to high-bandwidth control loops at the servo level, and knowledge-based control is introduced at higher levels in a control hierarchy in order to effectively monitor the robot performance and to tune the hard controllers at the servo level. Such a structure is expected to be superior to the use of soft control in servo-level control loops, as has been done in earlier applications of fuzzy control to robotic manipulators.

A schematic representation of the proposed control structure is shown in Figure 4.1. It consists of three control levels. The lowest level corresponds to the servo controllers at the robot joints (degrees of freedom). The joint servo loops are closed around a nonlinear feedback controller which decouples and linearises the robotic manipulator. The

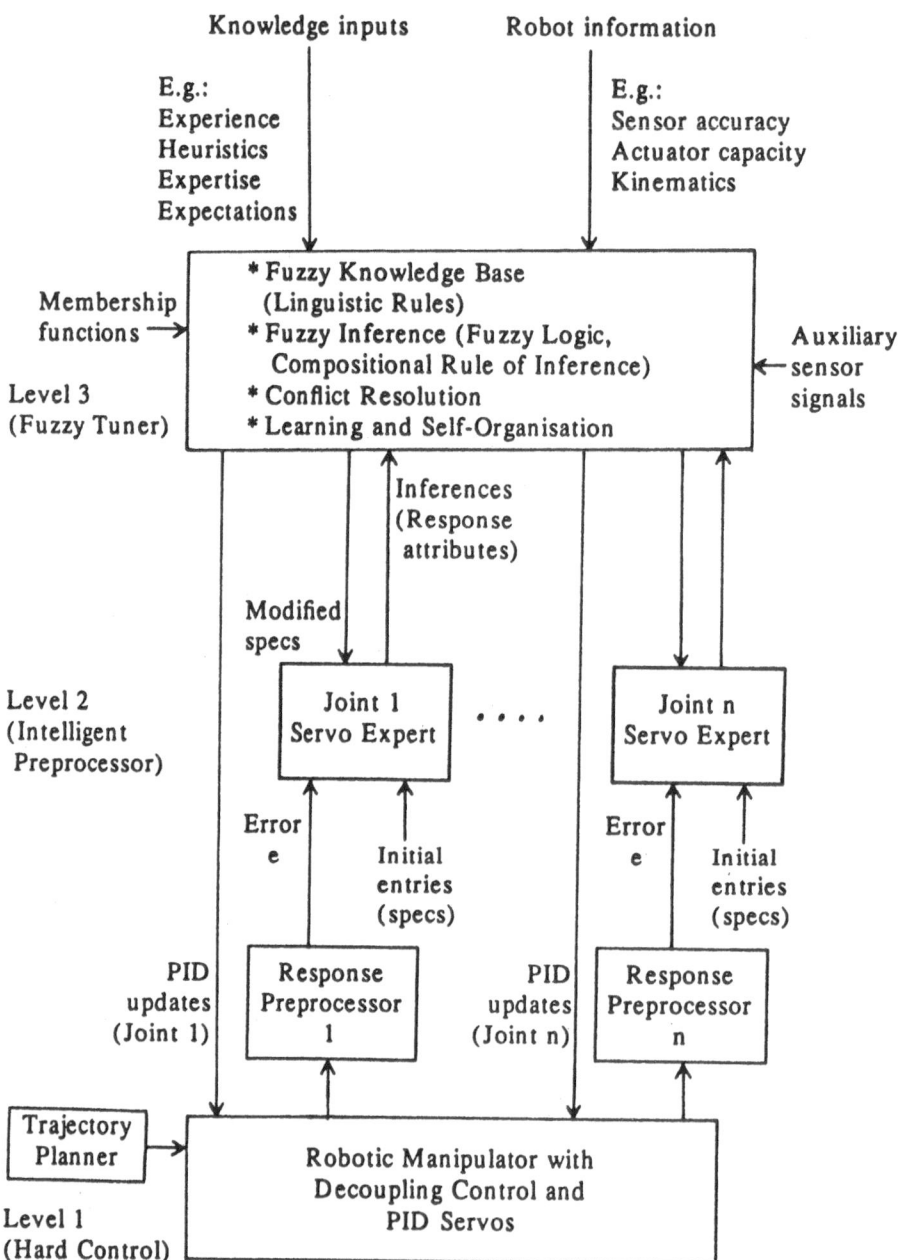

Figure 4.1 The Proposed Hierarchical Control Structure

robot trajectory is specified by the task. This desired trajectory is generated by the trajectory planner and resolved into desired joint trajectories, and these joint trajectories then form the reference inputs to the joint servos. The joint servos are assumed to be of the proportional-integral-derivative (PID) type. They are tuned in coordination, by a knowledge-based fuzzy controller at the third (highest) level of the control structure. The second level consists of a set of servo experts. These knowledge-based controllers are used to distribute the soft control activity among the degrees of freedom of the manipulator. There is one servo expert for each joint servo. It monitors the response of the particular degree of freedom, makes inferences as to the performance of that joint, with respect to a set of control specifications, and passes that information to the fuzzy controller, without regard to the remaining joints. The fuzzy controller uses this information, and possibly other forms of data and external knowledge, to determine appropriate modifications to the controller parameters in the servo level.

4.3 Decoupling and Servo Level

Simple linear servo control is known to be inadequate for transient and high-speed operation of robots (Horn and Raibert, 1978; Van Brussels and Vastmans, 1984; de Silva, 1984). Nevertheless, past experience of servo control in process applications is extensive, and this is true to some degree for industrial robots because servo control is the primary means of control in the present generation of commercial robots. This knowledge can be captured by a knowledge-based controller and used in servo tuning. But for this type of control to be effective, nonlinearities and dynamic coupling have to be compensated faster than the control bandwidth at the servo level. Therefore, a linearising and decoupling controller has to be implemented inside the joint servo loops.

The interest in nonlinear feedback control for linearising and decoupling complex processes is not new (Tokamaru and Iwai, 1971; Hemami and Camana, 1976). Of special interest here is the approach presented by Hewit and Burdess (1981).   An interpretation of their nonlinear feedback control scheme is shown in Figure 4.2. By following the procedures described in Chapter 1, a robotic manipulator may be modeled as:

$$\underline{M}(q)\ddot{q} \; = \; n(q,\dot{q}) + f(t) \tag{4.1}$$

where $\underline{M}$ is the nxn inertia matrix and f represents the vector of drive forces or torques. The number of degrees of freedom of the robot is n (typically 6) and the vector n represents nonlinear terms including coriolis and centrifugal forces, damping (say, coulomb friction), backlash, and gravitational forces. It should be clear that n is difficult to model, and $\underline{M}$, though containing nonlinear terms (trigonometric terms arising from the coordinate transformations), is relatively easy to model. If measurements of the joint forces (torques), joint positions, and joint accelerations are available, then the nonlinear term n can be computed using equation(4.1). This estimate of n is then fed back, as in Figure 4.2, to compensate for any unknown nonlinear effects. Furthermore, since an estimate of $\underline{M}$ is also available, it can be used in the forward path of the control loop as shown, to linearise and possibly decouple the robot dynamics. Even though the

$$\underline{M}(\underline{q})\ddot{\underline{q}} = \underline{n}(\underline{q},\dot{\underline{q}}) + \underline{f}(t)$$

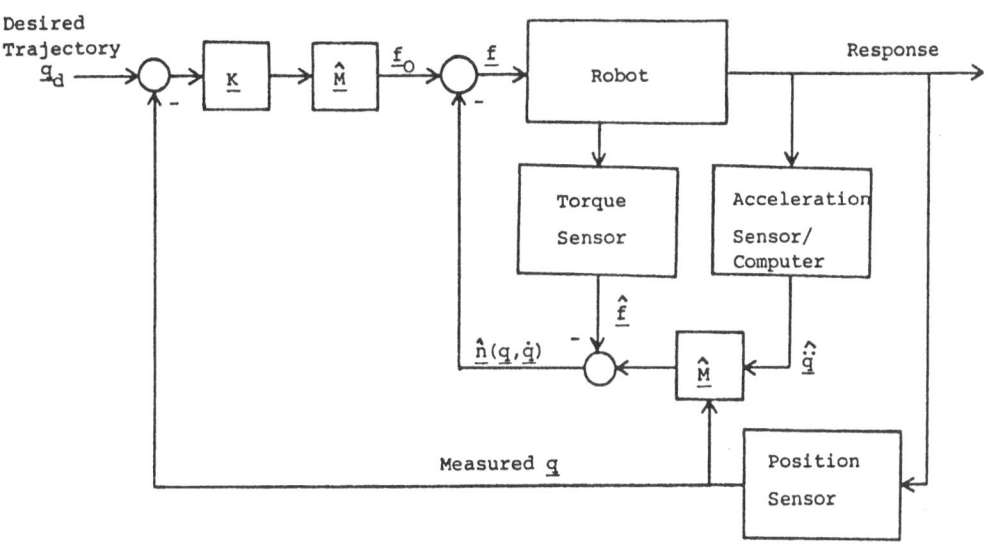

$$\underline{M}(\underline{q})\ddot{\underline{q}} \quad = \quad (\underline{n} - \hat{\underline{n}}) \quad + \quad \underline{f}_0$$

$$= \quad (\underline{n} - \hat{\underline{n}}) \quad + \quad \hat{\underline{M}}\ \underline{K}\ (\underline{q}_d - \underline{q})$$

If $\hat{\underline{M}} = \underline{M}$ and $\hat{\underline{n}} = \underline{n}$

$$\ddot{\underline{q}} = \underline{K}\ (\underline{q}_d - \underline{q})$$

Figure 4.2 An Active Nonlinear Feedback Control Scheme for Robots

method appears simple in theory, a practical implementation can be extremely difficult and even infeasible. Since differentiation of the joint velocity signal to obtain the joint acceleration can lead to well-known problems of noise, accelerometers for direct measurement of the signal would be needed. Furthermore, sensors for measuring drive forces (torques) at the manipulator joints would be also needed (de Silva, et. al., 1987). Since these measurements will not be exact, the estimation error of n could be high. For instance, an experiment conducted on a single joint by a group of researchers (including de Silva) failed to produce satisfactory estimates for n when single measurement samples of joint torque and joint acceleration were used to compute n in real time. Results could be improved through recursive estimation using several samples of data, but control bandwidth would be then affected due to the time needed for the estimation. Furthermore, even with only slight delays in the nonlinear feedback, large errors and stability problems can result, depending on how fast the nonlinear term n changes during operation. These problems can be avoided if n is modeled instead of computed through force and acceleration measurements. This model-based nonlinear feedback approach is what is proposed here for the lowest level of the control structure.

The proposed control structure for Level 1 of the hierarchy is thus as shown in Figure 4.3. Suppose that $\hat{n}$ is the model for for n. Then using the control law:

$$f = f_0 - \hat{n} \qquad\qquad (4.2)$$

where f_0 are the forcing signals to the joints, we get the system

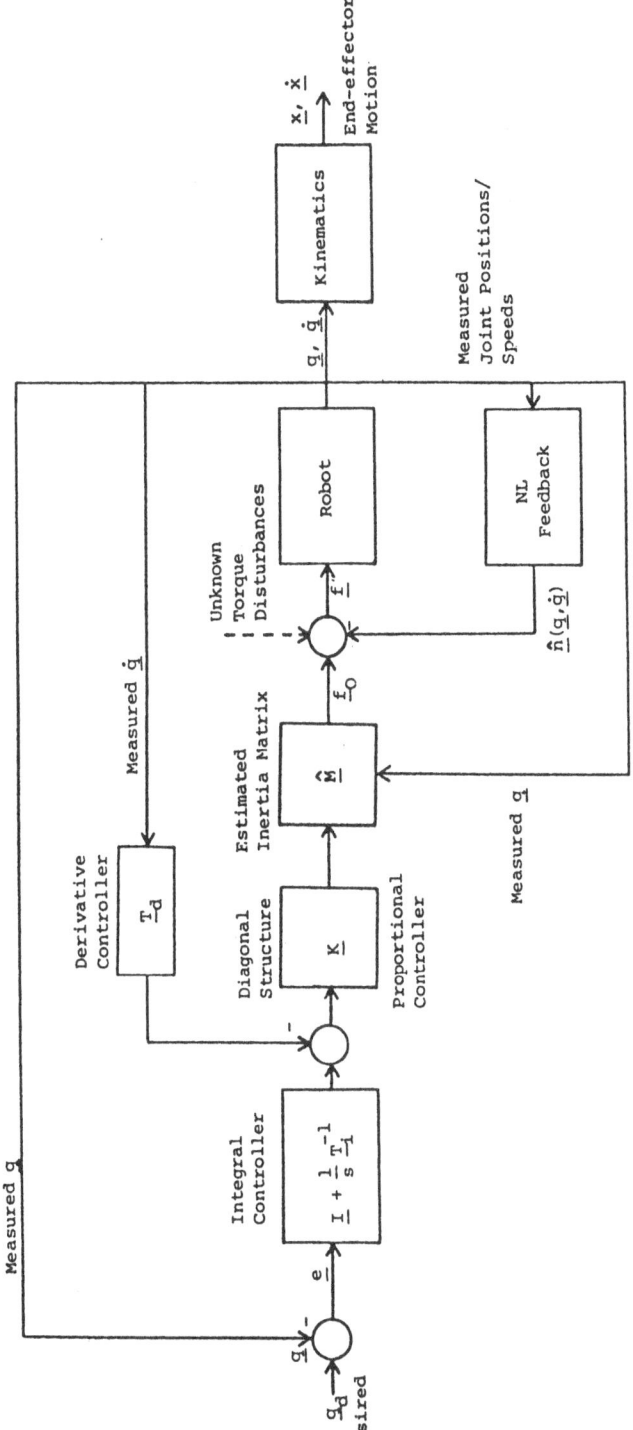

Figure 4.3 The Structure of the Level 1 Controller

equations:

$$\underline{M}\ddot{\mathbf{q}} = (\mathbf{n} - \hat{\mathbf{n}}) + \mathbf{f}_o \qquad (4.3)$$

The first term on the right hand side of equation (4.3) represents modeling error. Also, there will be unknown and usually unavoidable noise and disturbances in the system. For example, there will be torque disturbances in the drive unit and noise in the drive signal. Furthermore, as mentioned in Chapter 1, high-frequency backlash may be represented as an unknown disturbance. If all these unknown forcing components are denoted by $\mathbf{f}_e$, then equation (4.3) becomes:

$$\underline{M}\ddot{\mathbf{q}} = \mathbf{f}_e + \mathbf{f}_o \qquad (4.4)$$

Now suppose that the inertia matrix is modeled by $\hat{\underline{M}}(\mathbf{q})$. Also, suppose that the forcing signals are generated through a PID control law; thus:

$$\mathbf{f}_o = \hat{\underline{M}}\underline{K} \left[\mathbf{e} + \underline{I}_i^{-1} \int \mathbf{e} \, dt - \underline{I}_d \, \dot{\mathbf{q}} \right] \qquad (4.5)$$

where

$\mathbf{e} = \mathbf{q}_d - \mathbf{q}$ = error (correction) signal

$\underline{K}$ = decoupled structure (diagonal) matrix

$\underline{I}_i^{-1} = \underline{R}_i$ = positive-semidefinite and diagonal matrix of integral control rates (repeat rates)

$\underline{I}_d$ = positive-semidefinite and diagonal matrix of derivative control time constants

Consequently, the system may be represented by:

$$\hat{\underline{M}}\ddot{q} = \hat{f}_e + \hat{\underline{M}}\underline{K}[e + \underline{R}_i \; e \; dt - \underline{I}_d \dot{q}] \qquad (4.6)$$

where

$$\hat{f}_e = f_e + (\hat{\underline{M}} - \underline{M})\ddot{q} \qquad (4.7)$$

Note that all modeling errors and unknown forcing disturbances are included in the vector $\hat{f}_e$. Now, since $\hat{\underline{M}}$ is positive definite, an equivalent disturbance acceleration vector d can be defined as:

$$d = \hat{\underline{M}}^{-1}f_e \qquad (4.8)$$

Then, by premultiplying equation (4.7) by $\hat{\underline{M}}^{-1}$ we get the linear and noninteracting (decoupled) system:

$$\ddot{q} = \underline{K}[e + \underline{R}_i \int e \; dt - \underline{I}_d \dot{q}] + d \qquad (4.9)$$

The decoupled structure matrix is taken positive definite, with diagonal elements ω_j^2, where ω_j represents the modified natural frequencies of the manipulator joints. Also, $\underline{R}_i$ has diagonal elements r_{ij} and $\underline{I}_d$ has diagonal elements τ_{dj}. Hence the individual components of equation (4.9) are given by

$$\ddot{q}_j \;\; = \;\; \omega_j^2 \left(e_j + r_{ij} \int e_j \, dt \, - \, \tau_{dj} \, \dot{q}_j \right) \, + \; d_j \qquad\qquad (4.10)$$

$$j \; = 1, \, 2, \, ..., \, n$$

These parameters, ω_j, r_{ij} and τ_{dj} are the ones which are modified by the fuzzy controller during operation, if the response of robot is unsatisfactory. Essentially this is the way in which the modeling error and disturbance terms d_j are compensated for in this overall scheme of control. The rationale for this approach will be discussed, under knowledge-based control, in this chapter.

4.4 A Recursive Algorithm

The control structure which is developed here assumes the use of a high-bandwidth direct digital controller at the lowest level. In particular, a fast nonlinear feedback controller is needed for decoupling and linearising the nonlinear process. The nonlinear feedback control scheme which was described in the previous section relies on a real-time computation of the inertia matrix and modeled nonlinear terms. In the present section a recursive algorithm is developed for this purpose.

Recursive algorithms are available for computation of an input torque/force vector from a dynamic model of a robot. We shall extend the recursive Lagrangian formulation of robot dynamics (Hollerbach, 1980) to nonlinear feedback control. Consider the dynamic model:

$$\underline{M}(q)\ddot{q} \;\; = \;\; n(q,\dot{q}) + f(t) \qquad\qquad (4.11)$$

where, the vector $\mathbf{n}$ contains the nonlinear gravity terms and coriolis and centrifugal acceleration terms. At high speed operation of a robot, these nonlinear terms are known to cause the most difficulty in achieving the effectiveness of a joint servo (Horn and Raibert, 1978; Brussel and Vastmans, 1984; de Silva, 1984). The recursive algorithm given here undertakes to compensate for these nonlinear effects while decoupling the robot joints.

The algorithm calls for recursive computation of $\underline{M}$ and $\mathbf{n}$. The recursive Lagrange scheme given in Chapter 2 for computing the force/torque vector $\mathbf{f}$ can be used for this purpose. Initially, the homogeneous matrices $\underline{A}_j$ for link-to-link co-ordinate transformation and their first and second derivatives $d\underline{A}_j/dq_j$ and $d^2\underline{A}_j/dq_j^2$ are computed. Next the intermediate transformation matrices $\underline{W}_j$ and their first partial derivative with respect to q_j are computed as usual, using the following relations:

$$\underline{W}_0 = \underline{I}$$

$$\underline{W}_1 = \underline{A}_1$$

$$\delta\underline{W}_1/\delta q_1 = d\underline{A}_1/dq_1$$

$$\underline{W}_j = \underline{W}_{j-1}\underline{A}_j \qquad (4.12)$$

$$\delta\underline{W}_j/\delta q_j = \underline{W}_{j-1}d\underline{A}_j/dq_j \qquad (4.13)$$

$$j = 2, 3, ..., n$$

Computation of M(q)

The columns of the inertia matrix can be computed sequentially using a backward recursion (base to end effector) and a forward recursion (end effector to base). Note from equation(4.11) that if we set $\dot{q} = 0$ and the gravity vector $g = 0$, the nonlinear term $n(q,\dot{q})$ vanishes. Then, if we set $\ddot{q}$ equal to the unity vector whose i th element is unity, it follows that the force vector f becomes identically equal to the i th column of $\underline{M}$. In this manner, the recursive algorithm for force computation can be modified to compute the columns of $\underline{M}$. Specifically, the following three steps (backward recursion, forward recursion, and final non-recursive computation) are repeated for $j = 1, 2, ..., n$, to obtain the n columns of $\underline{M}(q)$:

$$\underline{P}_1 = \underline{P}_2 = ... = \underline{P}_{j-1} = \underline{0}$$

$$\underline{P}_j = \underline{W}_{j-1}d\underline{A}_j/dq_j$$
$$\underline{P}_i = \underline{P}_{i-1}\underline{A}_i \qquad (4.14)$$
$$i = j+1, j+2, ..., n$$

$$\underline{Q}_n = \underline{J}_n\underline{P}_n^T$$

$$\underline{Q}_i = \underline{J}_i\underline{P}_i^T + \underline{A}_{i+1}\underline{Q}_{i+1} \qquad (4.15)$$
$$i = n-1, n-2, ..., 1$$

Finally, the j th column of $\underline{M}(\mathbf{q})$ is obtained as

$$[\underline{M}]_{ij} = \text{tr}(\delta \underline{W}_i / \delta q_i \underline{Q}_i) \qquad (4.16)$$

$$i = 1, 2, ..., n$$

Conventional notations for the identity matrix, transpose, and trace have been used in the above equations.

Computation of n(q,q̇)

Note from equation(4.11) that if we set $\ddot{\mathbf{q}} = \mathbf{0}$, the force vector $\mathbf{f}$ becomes the negative of the $\mathbf{n}$ vector. This observation is employed in the following recursive relations for computing $\mathbf{n}$. First, backward and forward recursive relations are applied;

$$\dot{\underline{W}}_1 = d\underline{A}_1/dq_1 \dot{q}_1$$

$$\underline{P}_1 = d^2\underline{A}_1/dq_1^2 \dot{q}_1^2$$

$$\dot{\underline{W}}_i = \dot{\underline{W}}_{i-1}\underline{A}_i + \underline{W}_{i-1}d\underline{A}_i/dq_i \dot{q}_i$$

$$\underline{P}_i = \underline{P}_{i-1}\underline{A}_i + 2\dot{\underline{W}}_{i-1}d\underline{A}_i/dq_i \dot{q}_i + \underline{W}_{i-1}d^2\underline{A}_i/dq_i^2 \dot{q}_i^2$$

$$i = 2, 3, ..., n \qquad (4.17)$$

$$\underline{Q}_n = \underline{J}_n \underline{P}_n$$

$$\mathbf{c}_n = m_n \mathbf{r}_n$$

$$\underline{Q}_i = \underline{J}_i \underline{P}_i^T + \underline{A}_{i+1}\underline{Q}_{i+1}$$

$$\mathbf{c}_i = m_i \mathbf{r}_i + \underline{A}_{i+1}\mathbf{c}_{i+1} \qquad (4.18)$$

$$i = n-1, n-2, ..., 1$$

Note that m_i is the mass of the ith link and $\underline{J}_i$ is the moment of inertia matrix of that link expressed in and with respect to the body frame of the link. The vector position of the centroid of the ith link with respect to the body frame of the link is denoted by r_i. Finally, the elements of the vector $n(q,\dot{q})$ are computed using:

$$n_i(q,\dot{q}) = g^T \delta \underline{W}_i / \delta q_i \, c_i - \text{tr}(\delta \underline{W}_i / \delta q_i \, Q_i) \qquad (4.19)$$

By employing these recursive relations, the order of each vector computation is reduced from $O(n^4)$ to $O(n)$; this is a significant improvement in computational efficiency in comparison with a direct computation.

4.5 Servo Expert Level

There is a knowledge system for each degree of freedom of the manipulator. This knowledge system is termed a servo expert. Each servo expert will be implemented as a forward production system(FPS), having a knowledge base containing forward production rules, several data bases carrying data or context, and an inference engine for controlling the overall operation, as outlined in Chapter 1.

The servo expert monitors the error response of the particular degree of freedom of the robot. This monitored data forms the context for the servo expert. Inferences are made according to the rules in the knowledge base, independently without considering the response of the remaining joints. Typically, the error response is interpreted as a response from an independent second order process. Expertise for interpreting such responses should not be difficult to gain, particularly from process control practice. For example, a desirable response can be specified in terms of appropriate measures for the following characteristics:

1. Error
2. Speed of response or bandwidth
3. Stability

The fuzzy controller decides as to what actions should be taken, based on the inferences made by the servo experts. For example, if there appears a steady offset in the response, increasing the integral rate constant might be the solution; if the response appears slow, the natural frequency constant might have to be increased; an oscillatory response might be corrected by increasing the derivative time constant. Note that the specifications themselves might be updated (by the upper level fuzzy controller) depending on the operating conditions and performance requirements. If enough computing power is available, input-output data could be analysed and rules similar to Ziegler-Nichols tuning criteria, as commonly utilised in commercial knowledge-based servo tuners (Protuner, 1984) could be implemented in the knowledge base of the servo expert. Alternatively, knowledge engineering procedures which are followed in the development of expert systems could be followed to capture the knowledge of a group of experts in the servo control of robotic manipulators, to develop the rules *ab initio*.

If modeling errors are negligible, and if the control hardware and software in Level 1 are implemented perfectly, then individual joints should behave as independent simple oscillators with PID control. These are ideal goals, and in practice the presence of some dynamic coupling and nonlinear behaviour will be unavoidable. Furthermore, any disturbances, unknown influences, and software and hardware errors and delays will degrade the performance of the controller in Level 1. Hence inferences which are made by the individual servo experts will have to be further evaluated before making adjustments to the servos. This is one of the tasks which will be accomplished at the top level (fuzzy control level) of the hierarchy.

There are particular reasons for separating the servo expert level and the fuzzy control level. By attaching a knowledge-based controller to each degree of freedom of a robotic manipulator, we distribute intelligence among crucial locations of the robot. In this manner, localised monitoring can be done quickly and efficiently. Furthermore, priorities could be attached to the inferences which are made by the servo experts. By so doing, the fuzzy controller will be able to act quickly on troubled joints without waiting for inferences from other joints, particularly under emergency conditions. Under normal conditions, when the fuzzy controller has ample time to make its inferences, information from all servo experts at the lower level will be properly evaluated. In any event, it is necessary to delegate the final tuning decisions for individual joints to the fuzzy controller, where appropriate linguistic rules for servo control can be conveniently implemented. For example, the servo expert can supply to the fuzzy controller information on several attributes of the joint response without actually determining adjustments to the PID controller. In this sense, the servo experts act as intelligent preprocessors for the fuzzy tuner.

4.6 Fuzzy Control Level

It can be shown that by continuously adjusting (ideally at an infinite bandwidth) the servo parameters of a robotic manipulator, the joints can be made to respond like simple oscillators. This is the idea behind model referenced adaptive control of robots (Kornblugh, 1984). The disturbance terms d_j in equation(4.10) represent spill-over nonlinearities and dynamic coupling present in the robot system with a nonlinear feedback controller of the type shown in Figure 4.3. In theory, it is possible to completely eliminate d_j by adjusting the PID parameters. This is the aim of the fuzzy controller.

The primary objective of the fuzzy controller is to use its knowledge base (a set of linguistic fuzzy rules) to update the joint servo parameters, on the basis of the inferences from the servo experts, and other information which might be available. Generally, the fuzzy tuner is able to add, delete, or modify its own rule base and the rule bases of the servo experts. In addition, parameters of the nonlinear feedback controller could be adjusted, for example by using a measure for any dynamic coupling and nonlinearity which remains after nonlinear feedback. Furthermore, the performance specifications used by the servo experts might have to be updated during operation. It follows that the fuzzy controller performs learning and self-organisation functions as well as conflict resolution functions.

In particular, since the servo experts submit their inferences to the fuzzy controller, the decision as to whether to completely accept the recommendations which are made by a servo expert, or to modify or reject them on the basis of other information that might be available, is a task of conflict resolution. For example, if external sensors are used to monitor the response of the end effector, and that information is supplied to the fuzzy controller, an evaluation of the end-effector error can be used to determine a probable cause for that error. The fuzzy controller might be able to associate one or more joints with this error. As an example, suppose that the end effector trajectory has an offset error in the vertical direction. From kinematic considerations alone, it might be possible to determine which degrees of freedom are effective in correcting this error. If, for instance, there is a prismatic joint that is positioned vertically at the time, then, that joint might be the best choice for the corrective action. Available knowledge about joint sensors and actuators (e.g., resolutions, capacities) can also be used in decisions of this nature.

Information which is gathered by the fuzzy controller over a sufficiently long period of time can be used to modify performance parameters such as the response specifications which are used by the servo experts. This information gathering can be considered as learning. For example, by relaxing some specifications temporarily and by observing the resulting overall performance (at the end effector), it should be possible to determine whether or not the servo requirements are over-specified. The frequency of joint error monitoring, and of PID tuning, can also be changed through learning. Self organisation is a result of learning. In particular, modifications made to the rules and decision tables, on the basis of new information and experience, can be described as self organisation.

Inferences and actions which are made at the top level of the controller are equivalent to those which could have been made by a human expert. The associated knowledge can be expressed as a set of linguistic rules, and these form fuzzy information. The knowledge base of the fuzzy controller consists of fuzzy rules and fuzzy relations. Fuzzy information is stored as membership functions. The extension principle (See Chapter 1 and Chapter 3) is the primary means of extracting information from fuzzy relations. When nonfuzzy information, such as sensor readings, is available for a fuzzy quantity, the associated membership function must be available in the variable form so that the peak could be adjusted according to the available nonfuzzy information.

Tuning decisions for PID control parameters made by a human expert, arrived at on the basis of a set of attributes (e.g., oscillation amplitude, steady offset, convergence rate, divergence), can be expressed as a set of linguistic rules. These rules can be conveniently translated into fuzzy knowledge. In particular, statements relating attribute characteristics and tuning actions can be expressed as decision tables, with membership functions defined for the control actions. In this manner, tuning actions can be determined by the fuzzy controller using the attribute information on a joint response as provided by a servo expert in Level 2.

5. SYSTEM DEVELOPMENT

5.1 Introduction

To explore and illustrate the knowledge-based control structure discussed in Chapter 4, a control system was developed for a two degree-of-freedom robot. This consists of a robot simulator, two servo experts for the joints of the robot, and a fuzzy controller to generate the tuning commands for the PID controllers of the joints.

The system was developed on a SUN-3/50 workstation and transferred on to a high-speed SUN-3/160 for subsequent demonstrations. The robot simulator was a C program developed, compiled, and stored as a separate UNIX process. The servo experts were developed using the commercially available MUSE AI Toolkit (MUSE, 1987) separately. Rules and other unstructured programs of the servo experts were written in PopTalk language and the structured code of the knowledge-based controller was developed using the editor_tool facility of MUSE. The fuzzy controller was developed independently and was subsequently integrated with the servo experts in the form of a set of decision tables. Real-time communication among processes was achieved using UNIX Socket capability, Channel objects and Stream objects of MUSE, and additional interface programs were written in PopTalk.

The specific application discussed here does not cover all aspects of the proposed control structure in detail. The purpose here is to illustrate the typical steps which are followed in developing a knowledge-based control system of the proposed type, by using a relatively simple example, and to show the feasibility of the technique. In a practical application, the controller would be developed on a host computer such as the SUN workstation used in the present application, then tested using simulated data as in the present application, and finally loaded into an applications computer (target machine) for testing the controller with real data. This last step, which involves routine steps of prototyping a controller, is not undertaken here.

5.2 Servo Experts

Each joint has an associated servo expert. Since these servo experts have identical structures, the approach which has been taken is to first develop one sevo expert and, after testing it with data inputs, simply to duplicate it for the second joint of the robot. A servo expert is a knowledge system. It consists of a knowledge source, containing a rule system of the forward production type, and several data bases. It can interact with other data bases (including notice boards) and user-generated programs written in PopTalk language. The servo experts were developed using the MUSE AI Toolkit.

The MUSE AI Toolkit

A complete description of the MUSE AI Toolkit which was used may be found in the user manuals (MUSE, 1987). This introduction provides an overview of some of the characteristics of the toolkit, so giving continuity to this chapter.

MUSE is known to provide a quite flexible environment for developing knowledge-based applications. Due to this flexibility, a variety of application structures can be developed using several forms of knowledge representations. Particularly, either forward production systems (FPS) or backward chaining systems (BCS) may be used. What is relevant in the present application is that MUSE can be used to develop a knowledge system consisting of a knowledge base of forward production rules, several data bases, and an inference engine. Conveniently, this is the structure chosen for the proposed servo expert.

MUSE applications are developed by building structures called objects and then interconnecting these modules, subject to some structural constraints. An object contains data slots as well as instructions. In a conventional program, instructions and data structures are integrated together over the entire program. Hence even a small change to a data structure could make the program nonfunctional. In object-oriented programs, the data structure of any object can be modified without affecting the other objects in the program.

The structure of a MUSE *object* is shown in Figure 5.1. An object contains a set of slots, and these slots can hold one or more other objects. This results in a tree structure which terminates (at a leaf node) with slots containing unstructured PopTalk program modules or data values. An object can respond to messages. A response will depend on the *method* associated with the message for that particular type of object. Methods are programmed by the user in PopTalk language, at the stage of defining the schemas of the objects. An object has a *schema* associated with it. A schema is a data type definition, and it is the structural skeleton (template) from which any number of objects of the type of that particular

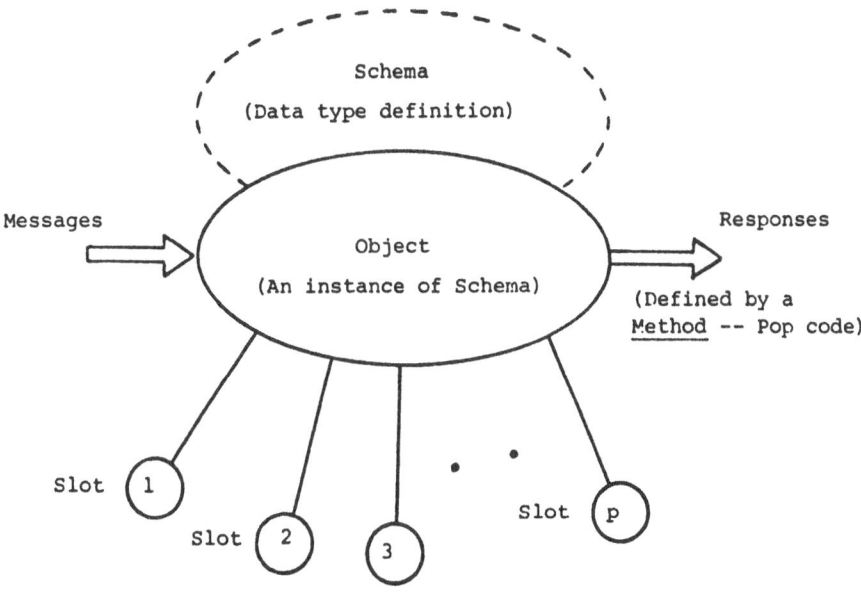

Figure 5.1 A Typical Object

schema, can be produced. In a MUSE application, knowledge is stored in a special type of object called a *knowledge source*. The structure of a knowledge source object is shown in Figure 5.2. It consists of a shared data base and a reasoning modules slot. The reasoning modules slot can contain one or more rulesets. A ruleset object has a private data base that is not visible outside the ruleset, and a *rules slot*. The rules slot has a collection of rule objects. Each rule object has a slot into which an unstructured PopTalk code that defines the particular rule in that ruleset of the knowledge base can be programmed. If the ruleset is of the FPR type, it corresponds to a forward production system (FPS). Backward chaining system (BCS) type rulesets are also available with MUSE, but are not used in the present application. A *notice board* is a data base which can store information that will be visible to other objects. The structure of a notice board is similar to that of a knowledge source, except that a notice board is a passive object and does not contain rulesets.

A typical Muse application is built by starting with a core object (handle) known as *system object* and developing and attaching knowledge sources and notice boards to the system object. This is schematically represented in Figure 5.3. Object manipulation may be done using the *structured editor* that is available with MUSE. Rules, methods, library programs, and other unstructured programs are developed in the PopTalk language using the EMACS editor. An *agenda* provides a scheduling mechanism for the execution of the knowledge sources in a given application.

Three other types of objects important in the present application are demons, data channels, and data streams. A *demon* provides a general

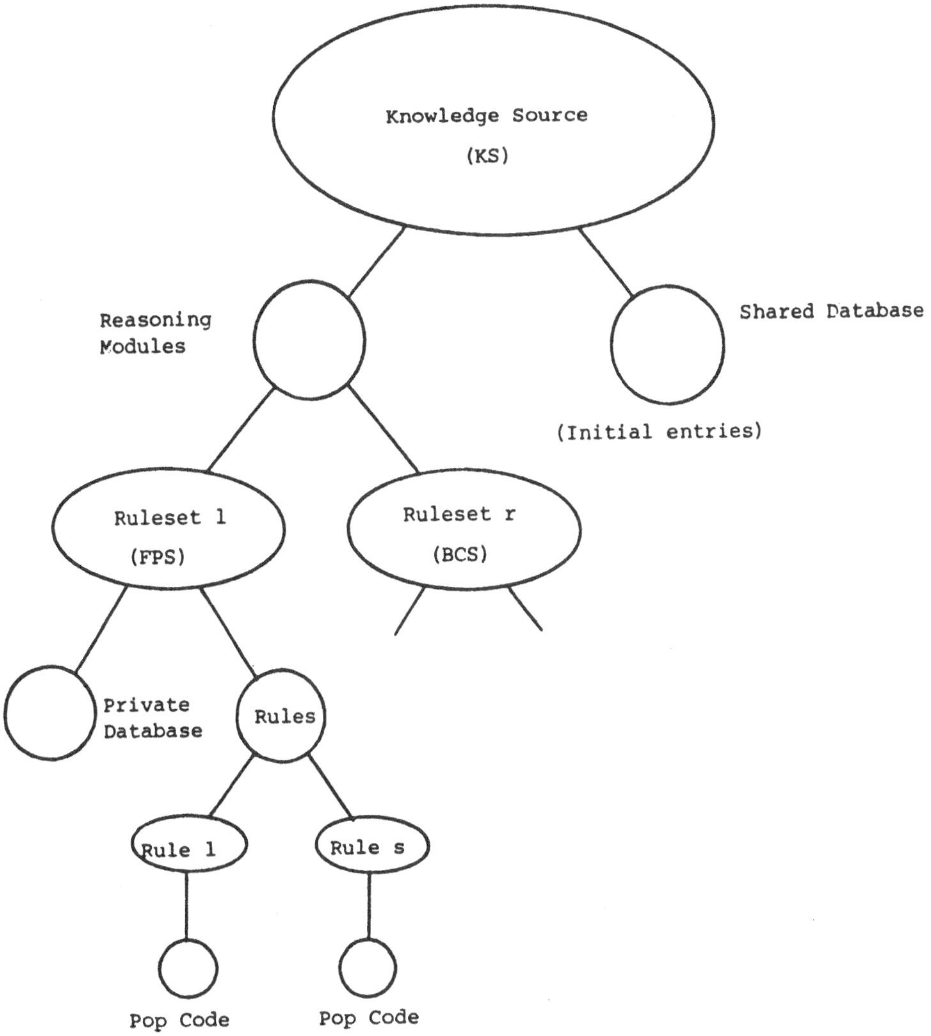

Figure 5.2   A Knowledge Source Object

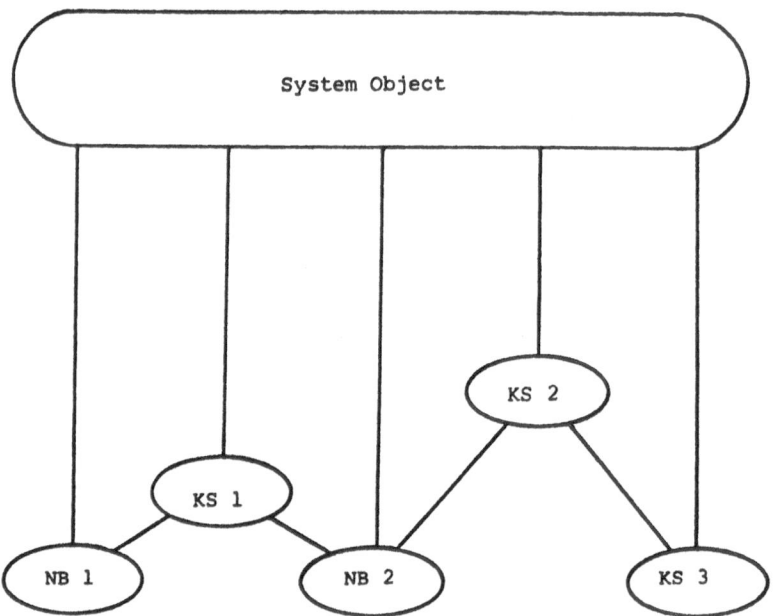

* Structured Editor

* EMACS (Unstructured) Editor

* PopTalk (Compile, Run) Pane

* Browser (Run-Time) Pane

Figure 5.3   A Typical MUSE Application

mechanism for monitoring objects. It can monitor creation, deletion, and updating of slots in an object, and can perform an action, as programmed in PopTalk, depending on the monitored condition. The action can be made either just prior to the change in the monitored object (pre-update demon) or after the monitored object has changed (post-update demon). A *data channel* is an object that can be interfaced to an external data source (e.g., a simulator) using the UNIX Socket mechanism. A channel object has an associated demon whose action can be programmed in PopTalk. When the channel receives an item of data, its demon is fired. A stream data channel is a data channel that can receive more than one item of data at a time. A *data stream* is an object that can be linked to a file outside the MUSE process. It can either read data from the file or write data into the file. In this manner channel objects and data stream objects can be used for linking MUSE processes with other external processes.

Servo Expert Development

The structure of a servo expert developed in the present application is shown in Figure 5.4. Each servo expert consists of a knowledge source, a notice board, and a PopTalk program for interfacing it to an external simulator through a UNIX socket. There is a servo expert for each degree of freedom (joint) of the robot. The notice board contains the specifications used in evaluating a joint response. The knowledge source contains a ruleset that has the rules carrying the necessary intelligence for evaluating a joint response. The inferences made on the joint response are stored as an object in the knowledge-source data base. These inferences are displayed during operation and also are used by the fuzzy controller for

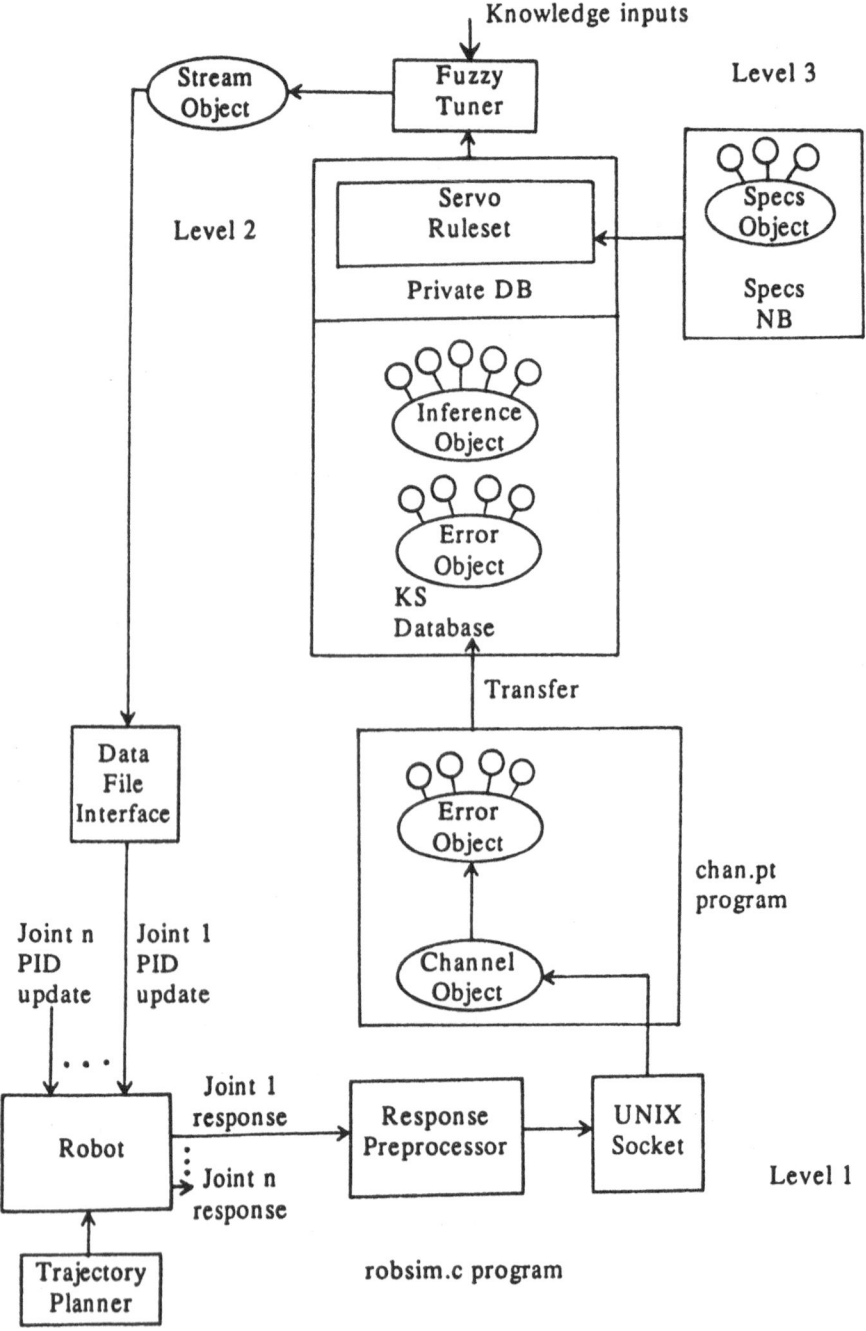

Figure 5.4 The Structure of a Servo Expert

making servo-tuning decisions at the top level of the control hierarchy. The PopTalk interface program associated with a servo expert creates a channel object for receiving joint error data from the robot simulator, through a UNIX socket interface. The program also creates an error object by taking three successive joint error values received by the channel object. The program is also responsible for the real-time transfer of these error objects to the data base of its servo expert.

In the present application three specifications are used for evaluating a joint:

1. An error tolerance (e_s)

2. An oscillation amplitude tolerance (a)

3. An acceptable rate of error convergence (λ)

Six rules are used in the rule base of each servo expert to test five attributes of a joint response. The five attributes considered are:

1. Accuracy

2. Oscillations

3. Speed of response (error convergence)

4. Divergence

5. Steady offset

The criteria which are used in the testing are illustrated in Figure 5.5. The first rule checks if the joint response is accurate. The second rule checks for unsatisfactory oscillations in the error response. The third rule monitors the speed of convergence of the joint error. The fourth rule

1. Accuracy ("okay"):

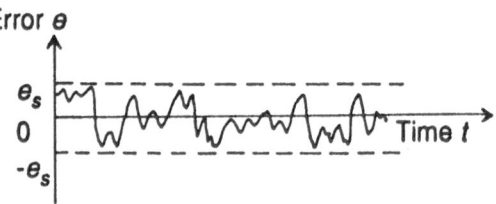

2. Oscillations ("absent"):

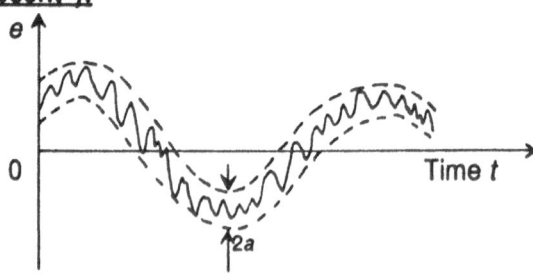

3. Decay ("acceptable"):

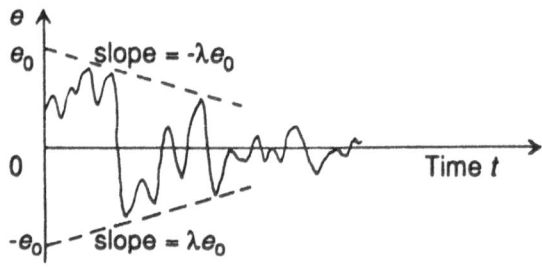

4. Divergence:

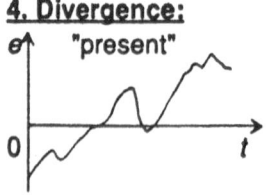
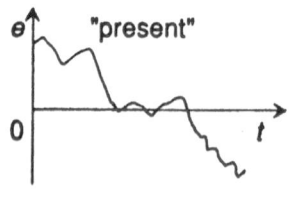
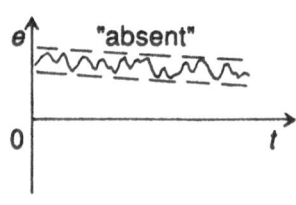

5. Offset:

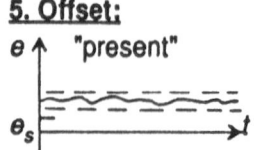
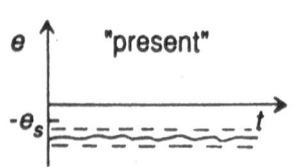
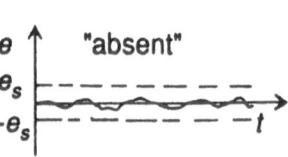

Figure 5.5 Some Attributes of a Joint Response

checks for steadily diverging unstable behaviour in the error response. The fifth rule checks for existence of a steady offset in the joint error. The final rule is needed for continuing the rule search when none of the previous five rules are fired in a given step, and this rule is the link for expanding the rule base.

Since a servo expert should not be expected to make inferences on the basis of a single data sample from a process, a *preprocessor* (filter) is used within the robot simulator to select the response error values for transmittal to the servo expert. First, an error monitoring period is chosen as an integral multiple of the sampling period of the robot simulator. This monitoring period determines the bandwidth of the knowledge-based controller. The preprocessor observes the joint error response over each monitoring period, and determines the alternate maximum and minimum values of error within successive error monitoring periods. These error values are sent sequentially to the channel object of the servo expert. The fuzzy controller has the ability to modify the monitoring speed (monitoring period) depending on the experience gained during operation.

The PopTalk source code of the servo expert program developed for the present application is listed in Appendix 1. The program was first tested by injecting error values, through an error notice board included during the development stage, and observing the inferences made by the reasoning mechanism. In every test, correct inferences were made by the servo expert. Admittedly, the rule base is not complete. The flexibility of a servo expert lies in the capacity to conveniently add new rules and delete unnecessary rules, as further experience is gained and expert advice is available, without having to modify other parts of the overall system. For example, a rule may be added to check for the presence of reset windup (integral windup) and to suppress the integral control action on that basis. After testing the first sevo expert, the error notice board was deleted and the servo expert was duplicated for the second joint of the robot. Subsequently, the new servo expert was modified to suit the relevant joint.

5.3 Robot Simulator

A program was written in C language, to simulate a robot. Since in the proposed control structure, nonlinear feedback is used at the servo level to decouple and linearise the robot, it is not necessary to program a nonlinear model of the robot. As discussed in Section 4.3, the response of the robot with decoupling control and PID servos can be represented by a set of simple oscillators with PID control and unknown disturbances. Hence the simplifying approach used here was to program this latter system, with acceleration-type disturbances provided by a set of user-defined functions. Various types of disturbances including random, pulse, step, periodic, and steady offset, can be easily included in this way. In an actual application, the recursive relations developed in Section 4.4 could be used to implement a high-bandwidth decoupling and linearising controller at the lowest level of the control structure. The present simulation takes for granted the effectiveness of such a control algorithm.

The structure of the robot simulator is shown in Figure 5.6. The inverse kinematics program, written in C, computes the desired joint trajectories corresponding to a desired end-effector tajectory programmed into it. The desired joint trajectories are then supplied as the

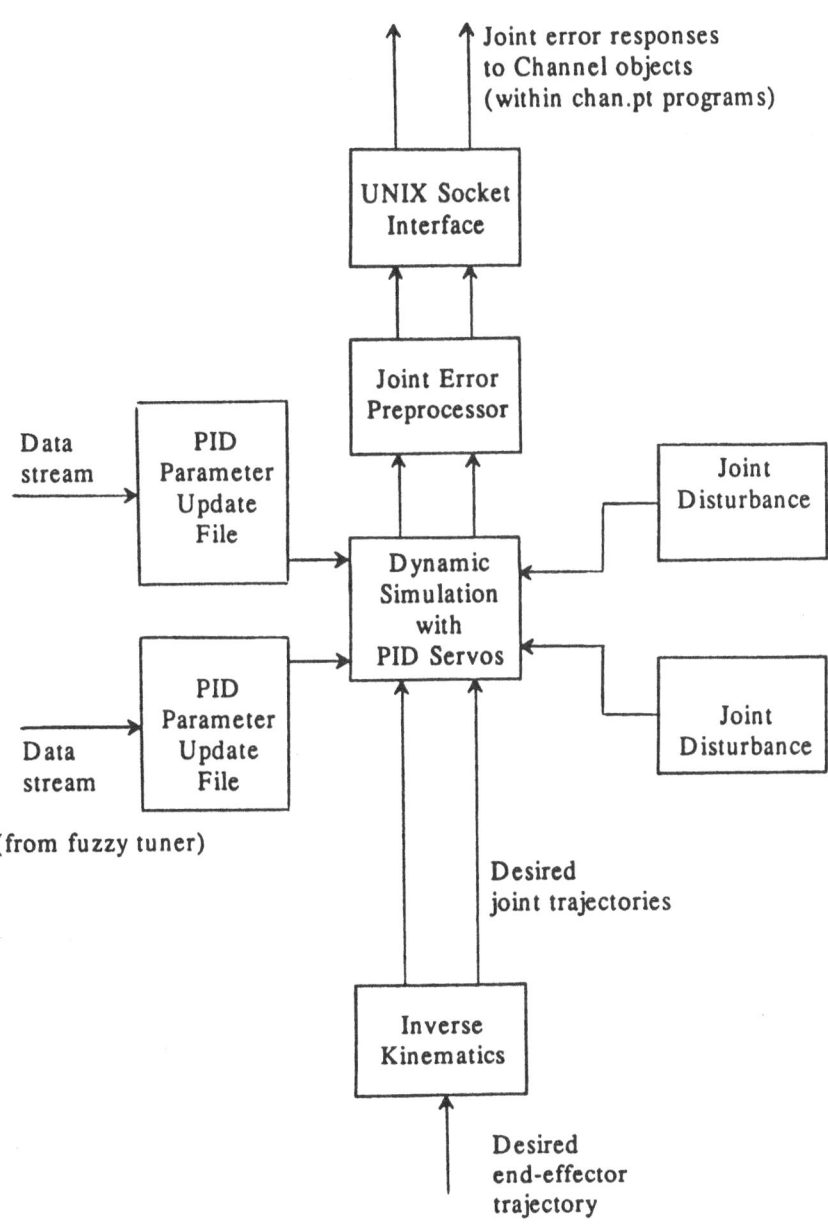

Figure 5.6 The Structure of the Robot Simulator

reference inputs to the joints of the simulated robot. The equations used in the inverse-kinematics computation are given below. Additional C program units were provided in the simulator for generating disturbance inputs to the joints. A preprocessor (a C program module) observes the actual joint error responses over each successive error monitoring period, and alternatively picks maximum value and minimum value (algebraic) of joint error within each period. These values are then sequentially transmitted to the channel object of the particular servo expert, through a UNIX socket interface. Each channel object is created by a separate PopTalk interface program. The same program constructs error objects from channel data, and transmits the error objects to an appropriate servo expert. The fuzzy controller sends PID parameter updates for each joint into separate data files. These data files are momentarily opened by the robot simulator during each monitoring period to read in the contents. These files are closed immediately after reading the data, so that they would be ready to receive further updates from the fuzzy controller. The robot-simulator C program including the preprocessor module, inverse kinematics unit and disturbance input units, and the PopTalk interface programs are listed in Appendix 1.

Inverse Kinematics

The inverse-kinematics problem for a robot is described in Chapter 2. In this section, the equations which are used in the computation of the inverse kinematics for the two-degree-of-freedom robot are given.

Consider the two-link robot shown in Figure 5.7, which executes planar motions in the cartesian frame (x , y). The position of the end effector is

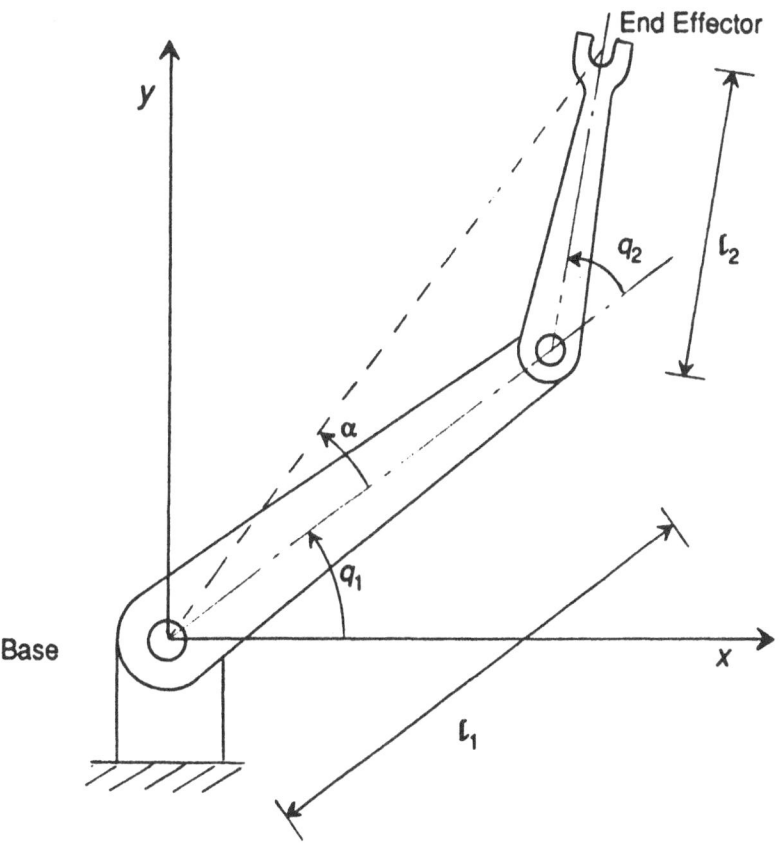

Figure 5.7   A Two-Link Robot

given by the vector:

$$r = \begin{bmatrix} r_x \\ r_y \end{bmatrix} \qquad (5.1)$$

It is seen that the direct-kinematics problem is governed by the two equations:

$$r_x = \ell_1 \cos q_1 + \ell_2 \cos (q_1 + q_2) \qquad (5.2)$$

$$r_y = \ell_1 \sin q_1 + \ell_2 \sin (q_1 + q_2) \qquad (5.3)$$

where ℓ_i are the link lengths and q_i are the joint coordinates, as shown in Figure 5.7.

The equations for inverse kinematics are obtained by simply applying the well-known cosine and sine formulae to the triangle formed by the links of the manipulator. Specifically, we get:

$$q_2 = \cos^{-1}[(r_x^2 + r_y^2 - \ell_1^2 - \ell_2^2)/(2\ell_1\ell_2)] \qquad (5.4)$$

$$q_1 = \tan^{-1} (r_y / r_x) - \alpha \qquad (5.5)$$

where angle α is as shown in the figure, and is given by:

$$\alpha = \sin^{-1}[\ell_2 \sin q_2 / \sqrt{r_x^2 + r_y^2}] \qquad (5.6)$$

In order to avoid nonunique solutions to the inverse-kinematics problem, the joint angles will be restricted to $0 < q_1 < 90°$ and $0 < q_2 < 180°$.

5.4 Fuzzy Controller

The next stage of the present application is the development of the fuzzy controller. First, expert knowledge on tuning a PID servo is expressed as a set of linguistic statements. Such statements (rules) will necessarily contain fuzzy quantities. Next, membership functions are established for the fuzzy quantities which are present in the fuzzy rules. Each fuzzy rule is then expressed as a fuzzy relation, in tabular form, by incorporating the membership functions into it through fuzzy logic connectives (e.g., **IF-THEN** implication and **OR**). The resulting set of tables forms the rule base of the fuzzy tuner. Rule matching can be accomplished by on-line application of the compositional rule of inference (see Chapter 3) to the fuzzy rule base, if so desired, using the inferences made by the servo experts as the context. This on-line application of the compositional rule is generally time consuming, and hence it may considerably reduce the speed of servo tuning. Since the universe of discourse of the servo-expert inferences (context) is discrete and finite in the present application, and since the universe of the tuning actions is also discrete and finite, both classes of universes having low cardinality, computational efficiency can be significantly improved by applying the compositional rule of inference off line. This preprocessing will generate a decision table for fuzzy tuning of each joint servo. Then, fuzzy decision tables developed in this manner can be directly integrated with the servo experts to perform PID tuning on line, during process operation. The steps of developing a fuzzy decision table for servo tuning are explained here. For brevity, the procedure is given in detail for one rule in the fuzzy rule set, and only the final results are given for the remaining rules.

Consider the set of linguistic statements given in Figure 5.8. These linguistic rules reflect the actions of a human expert in tuning a PID servo, by observing the error response of the sevo. Note the fuzzy quantities such as "highly oscillatory", "fast enough", and "small increase" which are present in these rules. It is clear that what is given in Figure 5.8 is a set of linguistic fuzzy rules for servo tuning. Of course, more rules can be added, and the resolution of various fuzzy quantities can be increased to improve tuning accuracy. But for the purpose of the present demonstration, this rule set is adequate. Now, using the notation given in Figure 5.9, we can express the fuzzy tuning rules in the condensed form shown in Figure 5.10.

Next we must develop membership functions for the fuzzy conditions OSC, RSP, DIV and OFF and for the fuzzy actions DP, DI and DD. Several methods for estimating such membership functions are given in Chapter 3. For our purposes, an approximate set of membership functions would suffice, based on intuition. Generally there is freedom to choose the resolution of each fuzzy quantity. In the present application, however, since inferences made by a servo expert are constrained to a discrete and finite set, the resolution of the condition variables would be fixed. For example, since the variable OSC can take one of three fuzzy values (OKY,

If the error response is not oscillatory *then* do not change the proportional *and* derivative control parameters;
or if the error response is moderately oscillatory *then* make a small decrease in the proportional gain *and* a small increase in the derivative time constant;
or if the error response is highly oscillatory *then* make a large decrease in the proportional gain *and* a large increase in the derivative time constant.

If the error response is fast enough *then* do not change the proportional *and* derivative control parameters;
or if the error response is not fast enough *then* make a small increase in the proportional gain *and* a small increase in the derivative time constant.

If the error response does not steadily diverge *then* do not change the proportional *and* integral *and* derivative control parameters;
or if the error response steadily diverges *then* slightly decrease the proportional gain *and* slightly decrease the integral rate *and* make a large increase in the derivative time constant.

If the error response does not have an offset *then* do not change the proportional *and* integral control parameters;
or if the error response has an offset *then* slightly increase the proportional gain *and* make a large increase in the integral rate.

Figure 5.8 Linguistic Fuzzy Rules for Servo Tuning

OSC = Oscillations in the error response
RSP = Speed of response (decay) of the error
DIV = Divergence of the error response
OFF = Offset in the error response

OKY = Satisfactory
MOD = Moderately unsatisfactory
HIG = Highly unsatisfactory
NOK = Unsatisfactory

DP = Change (relative) of the proportional gain
DI = Change (relative) of the integral rate
DD = Change (relative) of the derivative time constant

NH = Negative high (magnitude)
NL = Negative low (magnitude)
NC = No change
PL = Positive low
PH = Positive high

Figure 5.9 The Notation used for the Fuzzy Quantities

If	OSC	=	OKY	then	DP	=	NC
				and	DD	=	NC
or if	OSC	=	MOD	then	DP	=	NL
				and	DD	=	PL
or if	OSC	=	HIG	then	DP	=	NH
				and	DD	=	PH
If	RSP	=	OKY	then	DP	=	NC
				and	DD	=	NC
or if	RSP	=	NOK	then	DP	=	PL
				and	DD	=	PL
If	DIV	=	OKY	then	DP	=	NC
				and	DI	=	NC
				and	DD	=	NC
or if	DIV	=	NOK	then	DP	=	NL
				and	DI	=	NL
				and	DD	=	PH
If	OFF	=	OKY	then	DP	=	NC
				and	DI	=	NC
or if	OFF	=	NOK	then	DP	=	PL
				and	DI	=	PH

Figure 5.10 Condensed Form of the Linguistic Fuzzy Rules

MOD, and HIG), the membership function of OKY, for instance, should be assigned a universe having three elements only, each element being representative of one of the three fuzzy values. For action variables, however, this restriction of resolution is not necessary. For example, even though the action variable DP can assume one of four fuzzy values (NH, NL, NC, PL), the universe of, say NH, may contain more than four elements. However, in the present demonstration, in the interest of consistency and simplicity, we have decided to use the same restriction in resolution for both action variables and condition variables. Specifically, the cardinality of the universe of discourse of a fuzzy quantity is taken to be equal to the number of fuzzy values which can be assumed by the particular attribute. Accordingly, the universe of NH is assigned a cardinality of 4, and so on.

In an application of the present type, actual numerical values given to the elements in a universe of a fuzzy quantity are of significance only in a relative sense. But an appropriate physical meaning should be attached to each value. For instance, in defining DP, the numerical value -2 is chosen to represent a negative high change in the proportional gain. This choice is compatible with the choice of -1 to represent a negative low change. Also, such numerical values may be scaled and converted as necessary, to achieve physical and dimensional compatibility.

The chosen membership functions are shown in Tables 5.1 and 5.2. Note that, as desired, a membership grade of unity is assigned to the representative value of each fuzzy quantity. Fuzziness is introduced by assigning uniformly decreasing membership grades, starting with a low grade (0.2 or 0.1), to the remaining element values.

Next we shall show the development of a fuzzy relation table for a group

Table 5.1 Membership Functions of the Condition Variables

1. OSC:

	0	1	2
OKY	1.0	0.2	0.1
MOD	0.2	1.0	0.2
HIG	0.1	0.2	1.0

2. RSP:

	0	1
OKY	1.0	0.2
NOK	0.2	1.0

3. DIV:

	0	1
OKY	1.0	0.1
NOK	0.1	1.0

4. OFF:

	0	1
OKY	1.0	0.2
NOK	0.2	1.0

Table 5.2 Membership Functions of the Action Variables

1. DP:

	-2	-1	0	1
NH	1.0	0.2	0.1	0.0
NL	0.2	1.0	0.2	0.1
NC	0.1	0.2	1.0	0.2
PL	0.0	0.1	0.2	1.0

2. DI:

	-1	0	2
NL	1.0	0.2	0.0
NC	0.2	1.0	0.1
PH	0.0	0.1	1.0

3. DD:

	-1	0	1	2
NL	1.0	0.2	0.1	0.0
NC	0.2	1.0	0.2	0.1
PL	0.1	0.2	1.0	0.2
PH	0.0	0.1	0.2	1.0

of fuzzy rules. Table 5.3 illustrates the steps assciated with this development for the rules relating the condition OSC and the action DP. Note that there are three such rules in the rule base. For example, consider the rule:

If   OSC = OKY   **then**   DP = NC

To construct its fuzzy relation table R_1, we take the membership function of OKY from Table 5.1 and the membership function of NC from Table 5.2. Next, in view of the fact that fuzzy implication is a "min" operation on membership grades (see Chapter 3), we form the cartesian product space of these two membership function vectors and assign the lower value of each pair of membership grades to the corresponding location in the cartesian space. The remaining two relation tables R_2 and R_3 are obtained in a similar fashion. The three rules are connected by fuzzy **O R** connectives. Hence the composite relation table R is obtained by combining R_1, R_2, and R_3 through a "sup" operation: take the largest value of each triad of membership grades.

   The final step in the development of the fuzzy tuning controller is the establishment of the decision table. To explain this development, suppose that a servo expert infers the presence of moderate oscillations in the joint error response. Generally, this is a fuzzy inference, and its membership function is given by:

$$\mu_{OSC} = [0.2 \ \ 1 \ \ 0.2]$$

Table 5.3   Development of a Fuzzy Relation Table (for OSC --> DP)

R_1: If OSC = OKY then DP = NC

		DP			
		-2	-1	0	1
	0	0.1	0.2	1.0	0.2
OSC	1	0.1	0.2	0.2	0.2
	2	0.1	0.1	0.1	0.1

R_2: If OSC = MOD then DP = NL

		DP			
		-2	-1	0	1
	0	0.2	0.2	0.2	0.1
OSC	1	0.2	1.0	0.2	0.1
	2	0.2	0.2	0.2	0.1

R_3: If OSC = HIG then DP = NH

		DP			
		-2	-1	0	1
	0	0.1	0.1	0.1	0.0
OSC	1	0.2	0.2	0.1	0.0
	2	1.0	0.2	0.1	0.0

Composite R = R_1 V R_2 V R_3

		DP			
		-2	-1	0	1
	0	0.2	0.2	1.0	0.2
OSC	1	0.2	1.0	0.2	0.2
	2	1.0	0.2	0.2	0.1

Alternatively, the inference may be assumed crisp, with the membership function:

$$\mu_{OSC} = [0 \ 1 \ 0]$$

In either case, this context has to be matched with the rule base R obtained in Table 5.3. This is accomplished by applying the compositional rule of inference, as described in Chapter 3. Recall that this is a "sup of min" operation. Specifically, we compare the μ_{OSC} vector with each column of R, take the lower value in each pair of compared elements and then take the largest of the three elements thus obtained. It can easily be verified that, by this procedure we obtain:

$$\mu_{DP} = [0.2 \ 1.0 \ 0.2 \ 0.2]$$

as the membership function of the action on the proportional gain, corresponding to the inferred context. This result now has to be defuzzified, in order to obtain a crisp value for the tuning action. The centre of gravity method is used here. Specifically, we weight the elements in the universe (strictly, the support set) of DP using the membership grades of the action, and then take the average; thus:

$$((-2) \times 0.2 + (-1) \times 1.0 + 0 \times 0.2 + 1 \times 0.2)/4 = -0.3$$

This value is the entry for the DP action corresponding to the context

OSC = MOD, in the fuzzy decision table. The fuzzy decision table obtained by following these steps for every rule in the fuzzy rule base, is given in Table 5.4. It has been decided to take no action when the conditions are satisfactory, even though an optimal tuning strategy would suggest some other action under that condition. Also, the same decision table is used for both joints, since no discrimination is needed except for proper scaling. The relation used for updating a PID parameter is:

$$p_{new} = p_{old} + \Delta p \, (p_{max} - p_{min})/p_{sen} \qquad (5.7)$$

in which p denotes the PID parameter. The subscript "new" denotes the updated value and "old" denotes the previous value. The incremental action taken by the fuzzy controller is denoted by Δp. Upper and lower bounds for a parameter are denoted by the subscripts "max" and "min". A sensitivity parameter p_{sen} is also introduced for adjusting the sensitivity of tuning, when needed.

Table 5.4 Fuzzy Decision Table for a Joint Servo

Condition	Action		
	DP	DI	DD
OKY	0.0	0.0	0.0
OSC= MOD	-0.3	0.0	0.3
OSC= HIG	-0.5	0.0	0.5
RSP= NOK	0.15	0.0	0.3
DIV= NOK	-0.3	-0.3	0.5
OFF= NOK	0.15	0.6	0.0

6. PERFORMANCE EVALUATION

6.1 Introduction

This chapter presents some representative results obtained from the application described in Chapter 5. The performance of the knowledge-based controller is discussed on the basis of a set of simulation experiments, and some limitations of the present application are noted.

Several simulation experiments were carried out, using the application described in the previous chapter, in order to evaluate the performance of the proposed knowledge-based controller. In each experiment, a comparison was made with the robot performance under conventional control. The conventional controller was assumed to contain the same nonlinear feedback controller as in the knowledge-based controller, but the values of the PID servo parameters were not updated in the conventional control. In other words, the conventional-control results were obtained by bypassing the servo experts and the fuzzy tuner of the knowledge-based controller. It follows that the performance of the nonlinear feedback controller cannot be evaluated using the present experiments. But, as discussed in Chapter 4, errors of the nonlinear feedback controller can be represented by a set of disturbances injected into the joints of the robot. Specifically, the disturbance vector d in equation(4.9) accounts for possible poor performance of the nonlinear feedback controller. Joint disturbance inputs are provided by separate program units. Mutually dependent or independent disturbances could be injected into the joints in this manner.

6.2 Typical Operation

The robot control application was developed on a SUN Microsystems workstation (SUN-3/50). The simulation experiments could be demonstrated either on the same host machine or on a faster workstation (e.g., SUN-3/160). All programs including those for the robot simulator and knowledge-based control, were stored in the same directory (*smart_robot*). A typical simulation experiment is carried out as follows:

At the startup of the system, the *suntools* command is used to access the capability to interact with multiple and overlapping windows on the screen of the workstation. Then the *mouse* of the workstation can be used as a convenient user-interface device. The application directory (*smart_robot*) is accessed through the current system window. The MUSE reasoning framework is started by using the *editor_tool* command on this window. Now the MUSE window should appear. Next, a console window is opened, using the mouse. The robot simulator is run in a UNIX shell through this window; the simulator window.

At the start, the simulator code (C programs) is compiled through the simulator window, and the executable code is stored in a file named *robsim*. The data files which receive the PID updates from the fuzzy tuner, are initialised to zero during this step. Next the control is transferred to the MUSE window, through a mouse interaction. The knowledge-based control code is loaded and compiled through this window, using the PopTalk menu. At this stage, the screen of the workstation will appear as in Figure

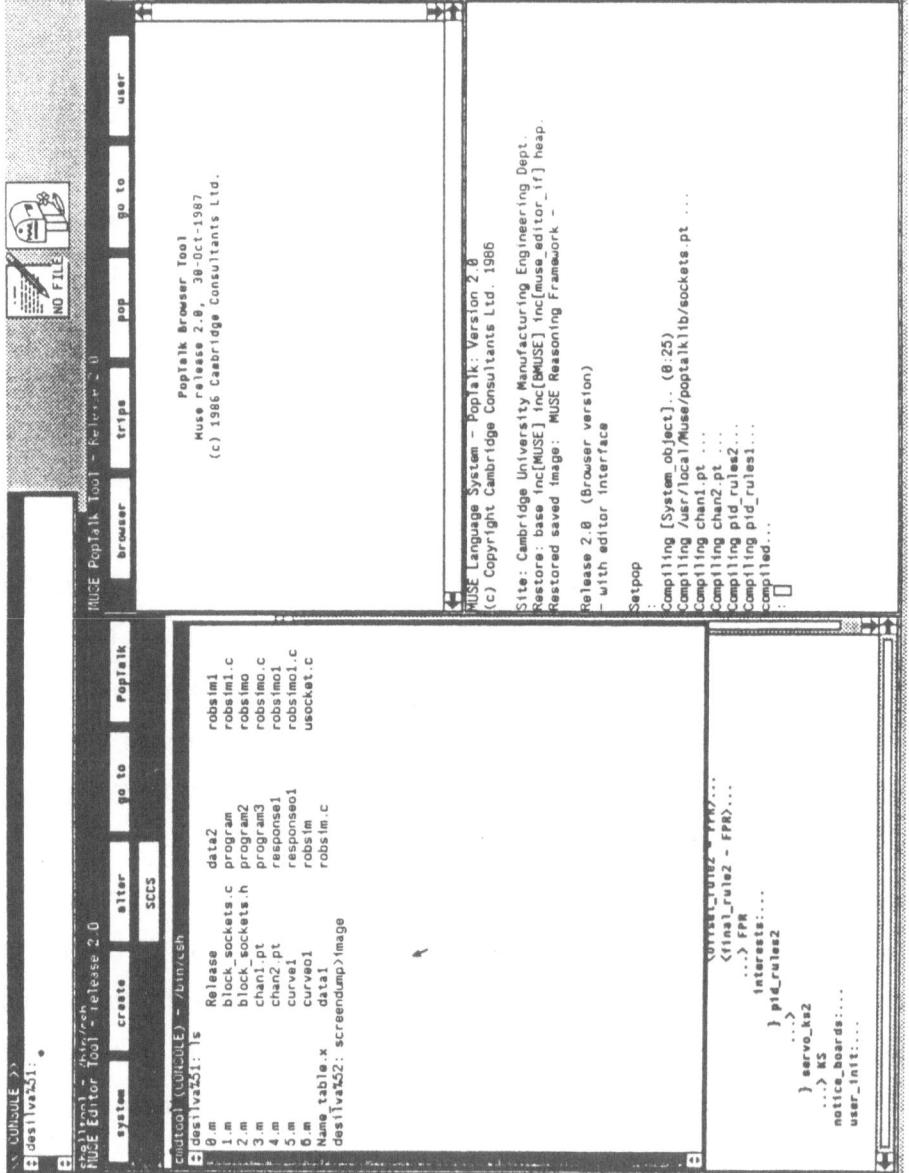

Figure 6.1 Typical Screen View at the Startup of the System

6.1. The left-hand window which hides the editor_tool window, is the simulator window. The right-hand bottom window is the PopTalk interaction window, and it is used for running the knowledge-based controller.

A simulation experiment is run by first executing the robot simulator using the *robsim* command on the simulator window, and then executing the knowledge-based controller by typing the *^run;* command on the PopTalk interaction window. During execution, the updated PID values will be printed on the simulator window, and the inferences from the servo experts will be displayed on the PopTalk interaction window. The appearance of the screen at the end of a typical simulation run is shown in Figure 6.2.

6.3 Single Joint Experiments

Initially, a joint of the robot was evaluated separately. The response of the joint was simulated for a step input, a ramp input and a sine input, under both conventional control and knowledge-based control. In these experiments, the following parameter values were used:

Natural frequency of the joint with no control (ω_0)   = 100.0 rad/s

Sampling period of the response simulation   (Δt)   = 2.0 ms

Integral rate of the servo controller (r_i)   = 1.0 s^{-1}

Derivative time constant of the servo controller (τ_d) = 5.0 ms

Bandwidth of the knowledge-based controller   = 62.5 Hz

Error convergence rate specification (λ)   = 100.0 s^{-1}

138

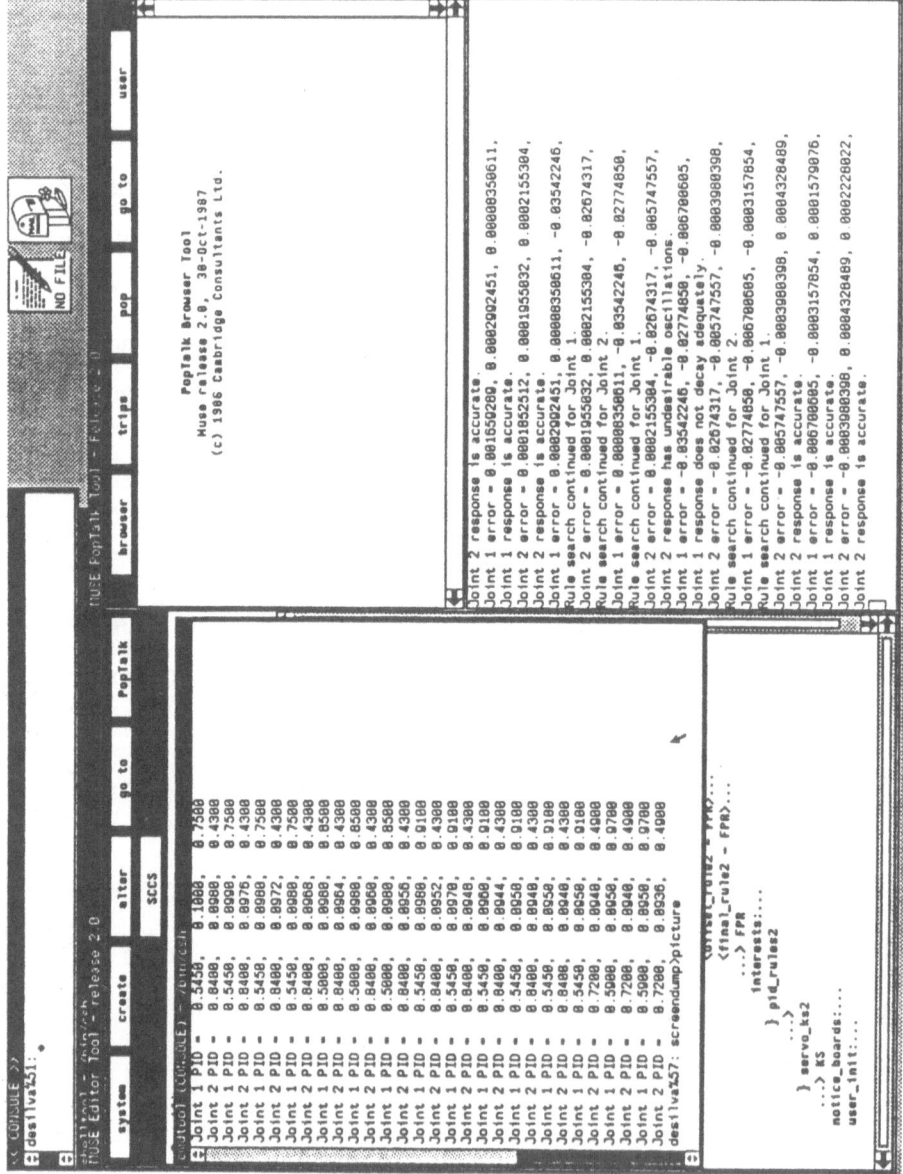

Figure 6.2 Typical Screen View at the End of a Simulation

Error tolerance specification (e_s)   = 0.05 rad

Oscillation amplitude tolerance specification (a)   = 0.025 rad

The servo parameter values given above are the values which were used for the conventional control simulations. They were also the starting values for the knowledge-based control simulations. The step input experiments were carried out over a duration of 0.1 sec., since both conventional control and knowledge-based control gave steady responses within this duration. Double this duration (0.2 sec.) was used for the ramp input experiments. A response duration of 1.0 sec. was used in the sine input experiments so that at least one full cycle of the input wave could be accommodated.

Results from the step-input tests are shown in Figures 6.3 and 6.4. With a knowledge-based controller of bandwidth 62.5 Hz,   satisfactory performance was obtained, but an overshoot was present, as shown in Figure 6.3(a). By increasing the bandwidth of the knowledge-based controller by a factor of four (to 250 Hz), it was possible to further improve the performance, completely eliminating the overshoot as shown in Figure 6.4(a).

In the ramp experiments, a ramp that reaches a unity magnitude at 0.2 sec. was used as the input, to be consistent with the step input tests. The results obtained are shown in Figure 6.5. The ramp response under knowledge-based control, as shown in Figure 6.5(a), is quite satisfactory. The knowledge-based controller appears to quickly tune the servo parameters so that accurate tracking of the input ramp is achieved very rapidly. Under conventional control, the ramp response is not as accurate,

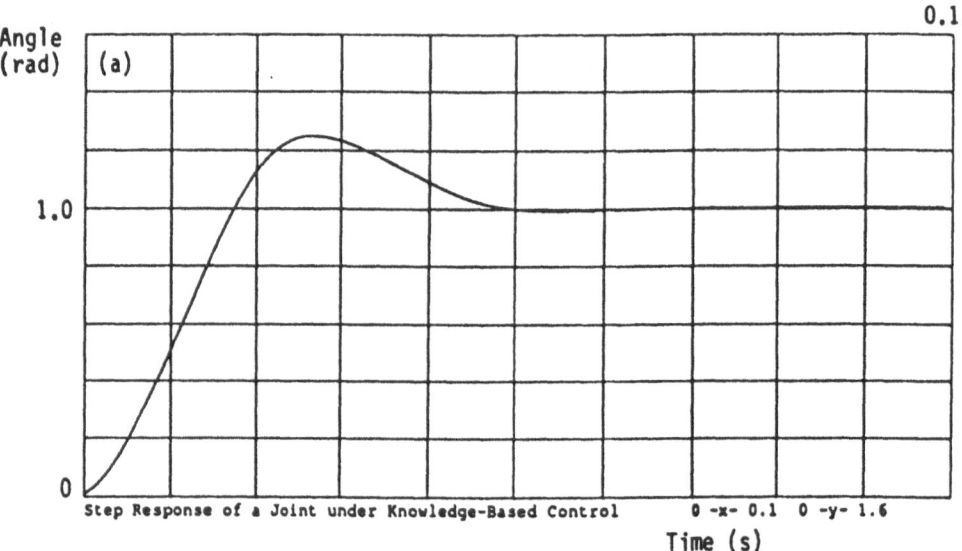

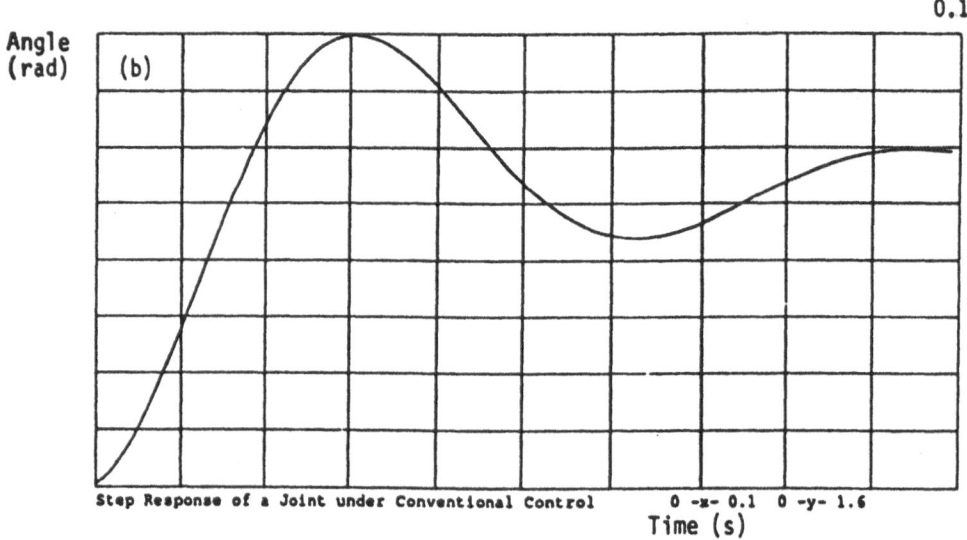

Figure 6.3   The Step Response of a Joint of the Robot
           (a) Under Low-Bandwidth Knowledge-Based Control
           (b) Under Conventional Control

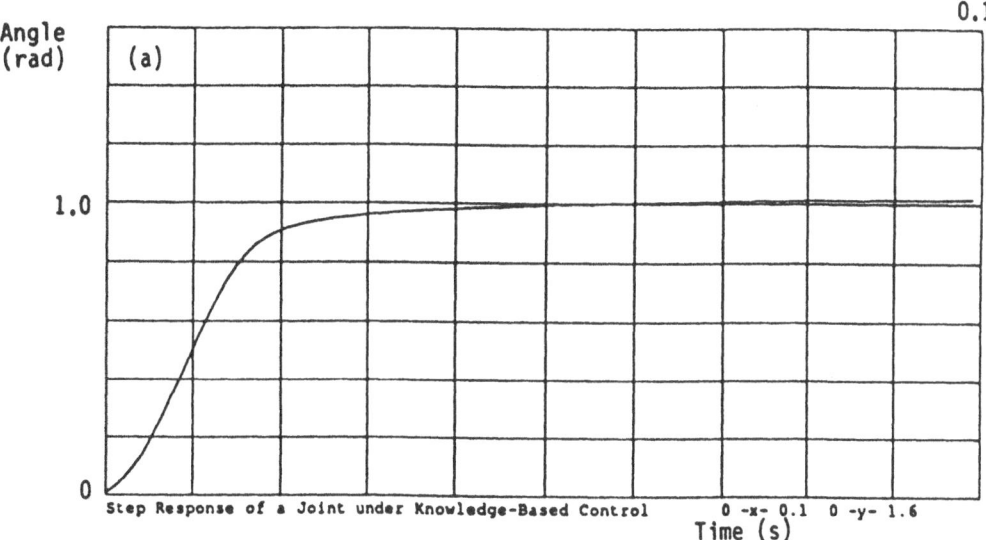

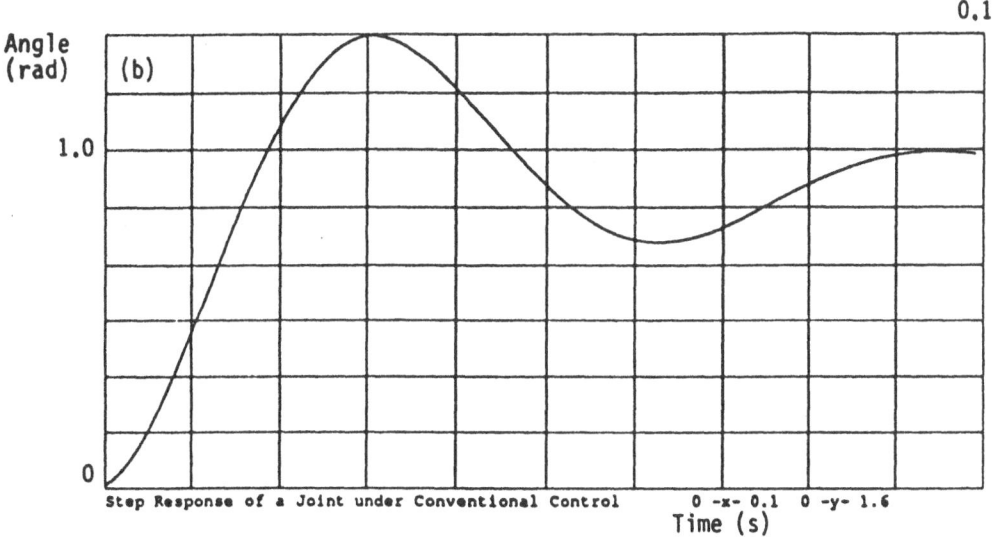

Figure 6.4   The Step Response of a Joint of the Robot
(a) Under High-Bandwidth Knowledge-Based Control
(b) Under Conventional Control

0.2

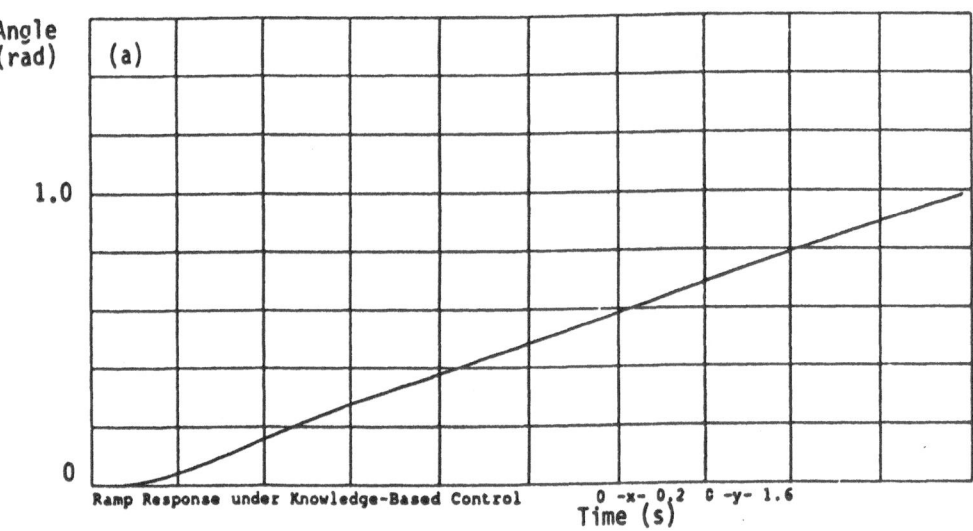

Angle
(rad)

(a)

1.0

0

Ramp Response under Knowledge-Based Control

Time (s)

0 -x- 0.2   0 -y- 1.6

0.2

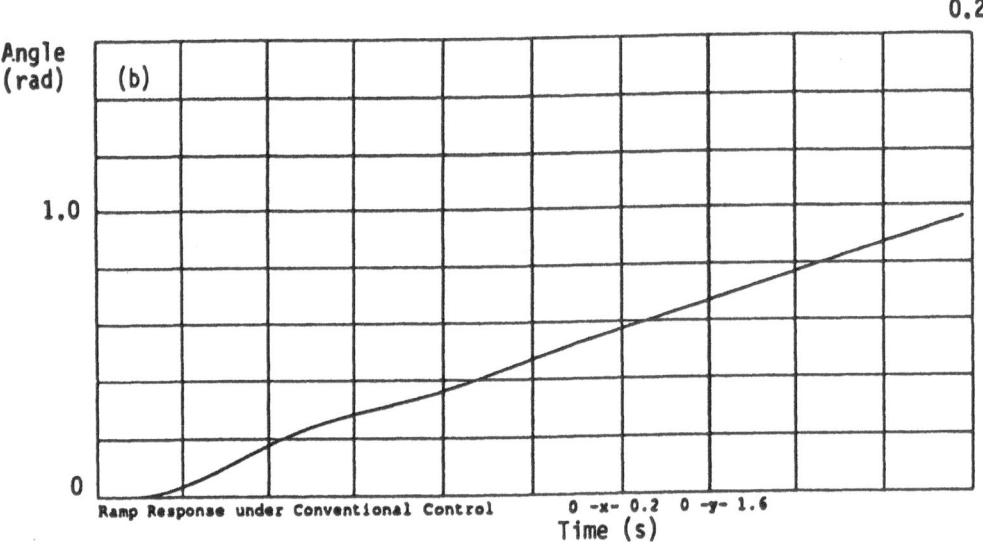

Angle
(rad)

(b)

1.0

0

Ramp Response under Conventional Control

Time (s)

0 -x- 0.2   0 -y- 1.6

Figure 6.5   The Ramp Response of a Joint of the Robot
(a) Under Knowledge-Based Control
(b) Under Conventional Control

as shown in Figure 6.5(b). In particular, the response lagged behind the input ramp throughout the duration; and furthermore a more oscillatory behaviour was noticed. On the other hand, the knowledge-based controller has identified the steady offset and has used its integral control action to virtually eliminate this offset.

Results obtained by applying a sine input are shown in Figures 6.6 and 6.7. With the knowledge-based controller, the response shown in Figure 6.6(b) was obtained for the sinusoidal input of unity amplitude shown in Figures 6.6(a) and 6.7(a). When the simulation experiment was repeated under conventional control, the response shown in Figure 6.7(b) was obtained. It is seen that a significant improvement in the performance has been achieved through the proposed knowledge-based control method.

6.4 Seam Tracking Experiments

A robot tracking a right-angular path was simulated, to demonstrate the performance of the proposed controller in seam-tracking tasks. Several considerations have to be taken into account in desgning experiments of this type. For example, a suitable time trajectory for tracking the path has to be used. Two types of trajectory were used in these simulations. First, uniformly accelerating and decelerating trajectories of equal duration, without any uniform speed segments, were used. This would be the case if the capacity of the joint actuators were such that the maximum operating speed is not reached during the task duration, and the task has to be completed in minimum time. It was assumed that the robot comes to rest at the corner of the right angle at a moderate deceleration, before changing direction. In the second type of trajectory, the end effector attempts to

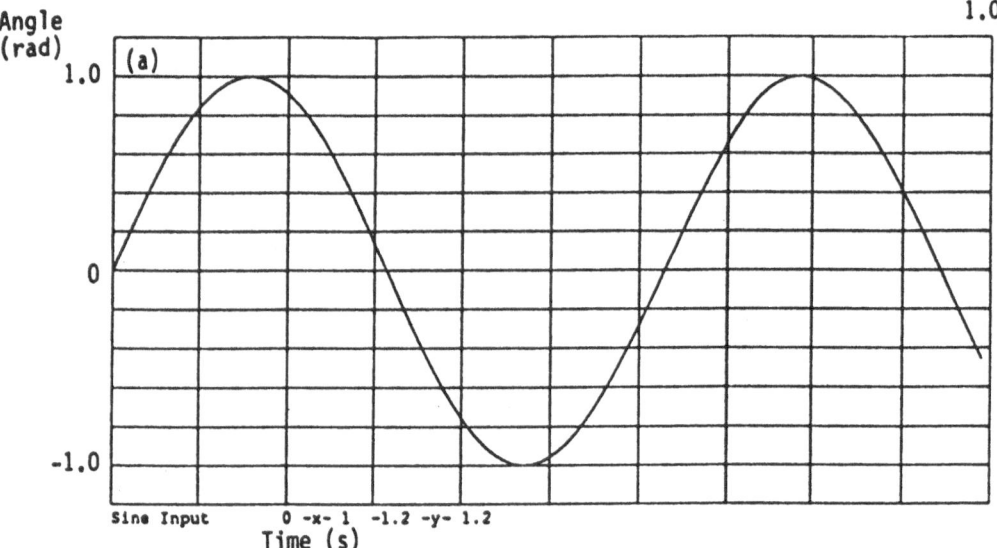

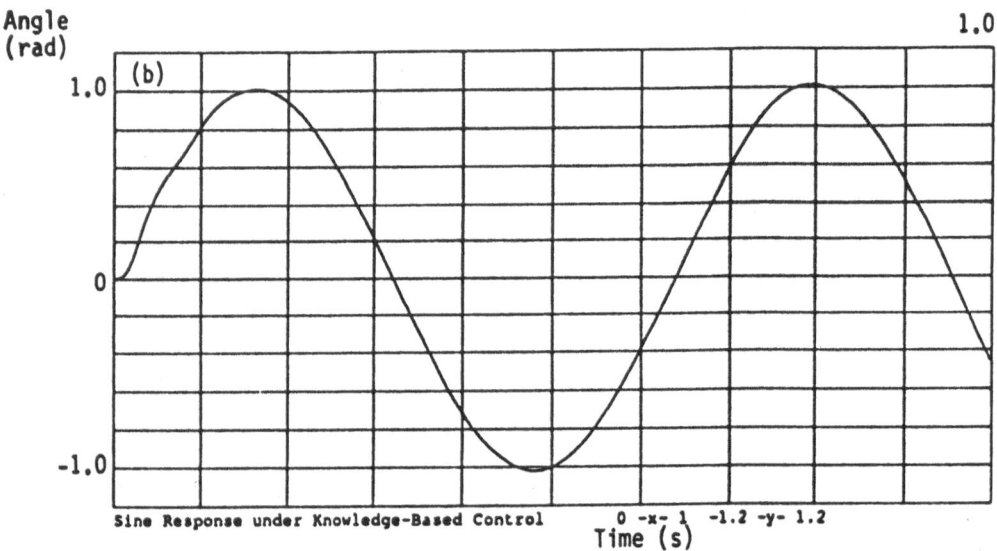

Figure 6.6   The Sine Response of a Joint of the Robot
              Under Knowledge-Based Control
              (a) The Input
              (b) The Response

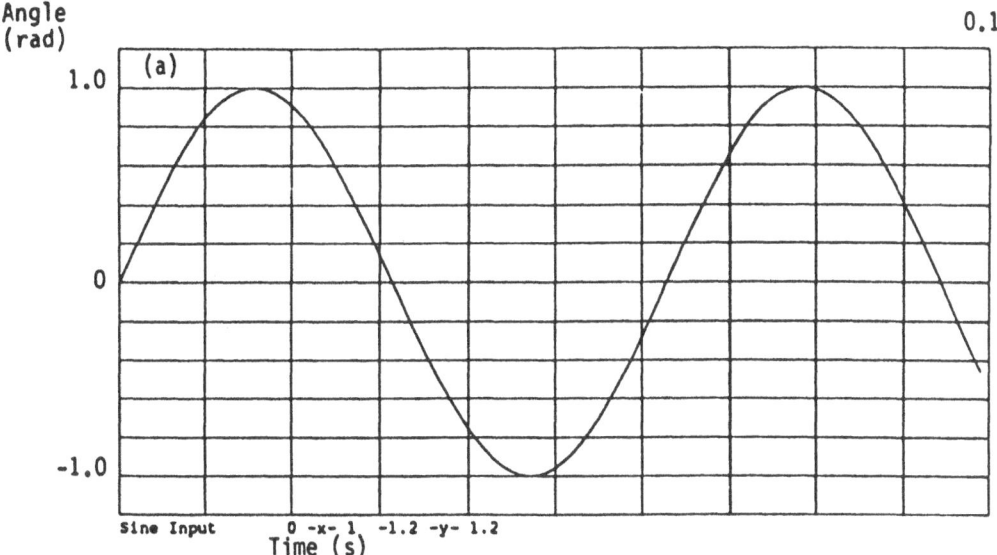

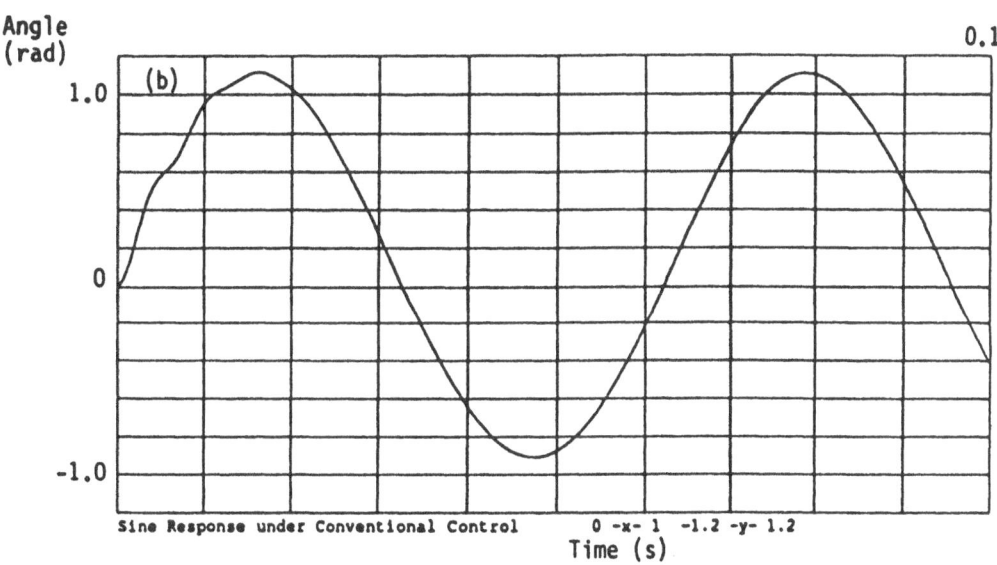

Figure 6.7   The Sine Response of a Joint of the Robot
Under Conventional Control
(a) The Input
(b) The Response

negotiate the corner much faster. Specifically, in the neighbourhood of the corner, the end effector is rapidly decelerated so that it comes to rest at the corner. Then the end effector is rapidly accelerated in the orthogonal direction, to its full speed. Uniform speed segments do not present such serious control problems as do accelerating or decelerating segments. For this reason, uniform speed segments were not included in the seam tracking experiments described here.

Another important consideration lies in the general nature of the robotic task. In some trajectory-following tasks such as arc welding and spray painting, it is difficult to start the robot under ideal operating conditions. Hence, an initial position error would be present. But, subsequently, uniform operating conditions would be attained. Tasks of this category were simulated by injecting position errors to both joints of the robot at the beginning of each task. In some other types of task, the pay load itself might change during operation. Pick-and-place operations, assembly tasks, and interactive tasks for multiple robots, or for a robot and at least one other machine tool (say, in a flexible manufacturing cell), would fall into this category. The robot might not stop while the condition change is taking place. Operation under an unsatisfactory nonlinear feedback controller can also lead to a robot behaviour similar to what would be present under these conditions. Robot operation under such conditions was simulated by injecting acceleration pulses to disturb the joints of the robot at different time points during the operation.

The following parameter values were used in the seam tracking experiments:

Natural frequency of Joint 1   with no control (ω_1)   = 100.0 rad/s

Natural frequency of Joint 2   with no control (ω_2)   = 150.0 rad/s

Sampling period of the response simulation   (Δt)   = 8.0 ms

Integral rate of Servo Controller 1 (r_{i1})   = 0.0 s⁻¹

Integral rate of Servo Controller 2 (r_{i2})   = 0.0 s⁻¹

Derivative time constant of Servo Controller 1 (τ_{d1})   = 0.1 ms

Derivative time constant of Servo Controller 2 (τ_{d2})   = 0.1 ms

Bandwidth of the knowledge-based controller   = 62.5 Hz

Error convergence rate specification (λ)   = 100.0 s⁻¹

Error tolerance specification for Joint 1   (e_{s1})   = 0.01 rad

Error tolerance specification for Joint 2   (e_{s2})   = 0.01 rad

Oscillation amplitude tolerance for Joint 1   (a_1)   = 0.005 rad

Oscillation amplitude tolerance for Joint 2   (a_2)   = 0.005 rad

In each experiment, a right-angular path of dimensions 1.0 m x 1.0 m was tracked in 4.0 sec. by a robot whose base link was 2.0 m long and whose end-effector link was 1.0 m long.

The robot was started with errors of magnitude 0.05 rad at the two joints. Note that under the worst circumstances, this can result in an initial position error of 15.0 cm. The end effector was uniformly accelerated from rest, decelerated to rest at the corner, accelerated again in the orthogonal direction, and decelerated to rest at the end of the path, during four successive periods of 1.0 sec. along the trajectory. Figure 6.8(a) shows the response of the robot under knowledge-based control. The controller first corrects the initial error and then maintains the accuracy of the response thereafter. The response sown in Figure 6.8(b) indicates that the conventional controller was unable to eliminate the position error. It should be mentioned here that better performance could have been obtained from the conventional controller by tuning its servo parameters. By the same token, the performance of the knowledge-based controller also could have been further improved by properly selecting the initial parameter values for the joint servos. In the present simulations, servo parameter values used in the conventional controller are the starting values of the knowledge-based controller. It is not always easy or straightforward to select proper parameter values for a controller; the knowledge-based controller performs well even with an unsatisfactory set of initial parameter values while the conventional controller does not.

Next the simulation was repeated, this time with acceleration disturbances of magnitude 0.05 rad/s^2 injected to Joint 1 (base joint) of the robot at times 0.8 sec., 1.6 sec., 2.4 sec., and 3.2 sec., and acceleration disturbances of magnitude 0.1 rad/s^2 injected independently to Joint 2 (end effector joint) at times 0.6 sec., 1.4 sec., 2.2 sec., and 3.4 sec. over the 4.0 sec. duration of the trajectory. Note that a joint acceleration of magnitude 0.1 rad/s^2 can create a force of 0.3 N on a 1.0 kg mass at the end effector. The results of this simulation are shown in Figure 6.9. Again, the knowledge-based controller has been able to compensate for the effects of the disturbances much better than has the conventional controller.

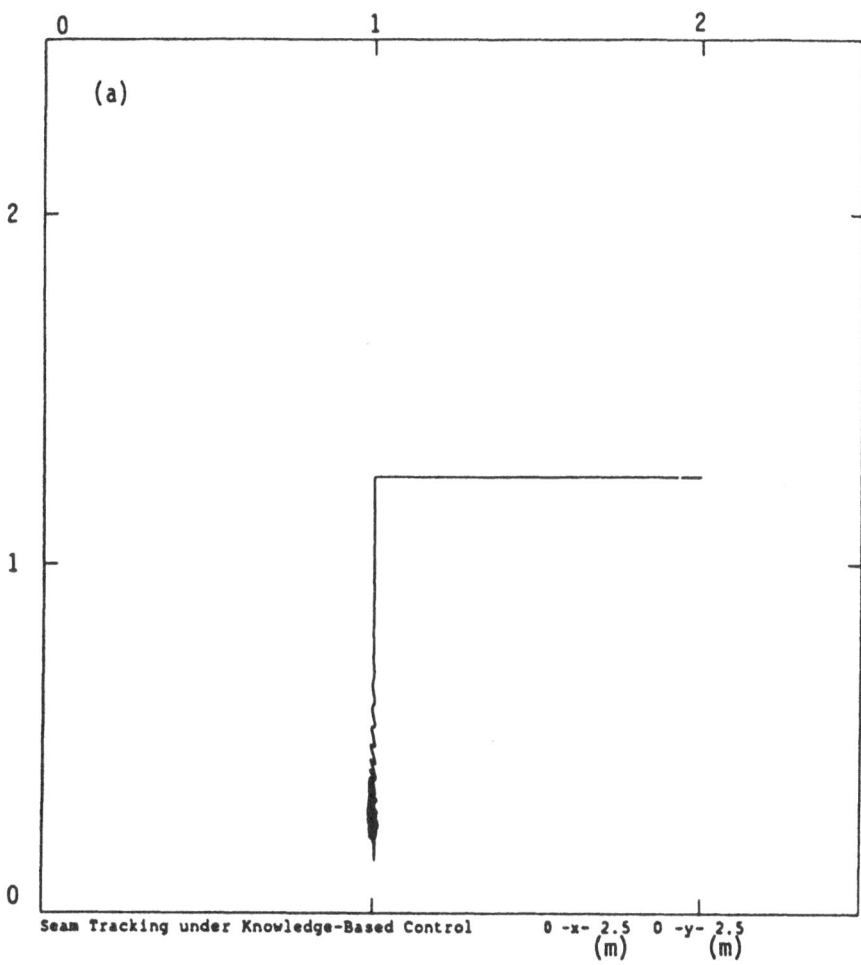

Seam Tracking under Knowledge-Based Control   0 -x- 2.5   0 -y- 2.5
                                                   (m)   (m)

Figure 6.8   Seam Tracking with an Initial Position Error
              (a) Under Knowledge-Based Control
              (b) Under Conventional Control

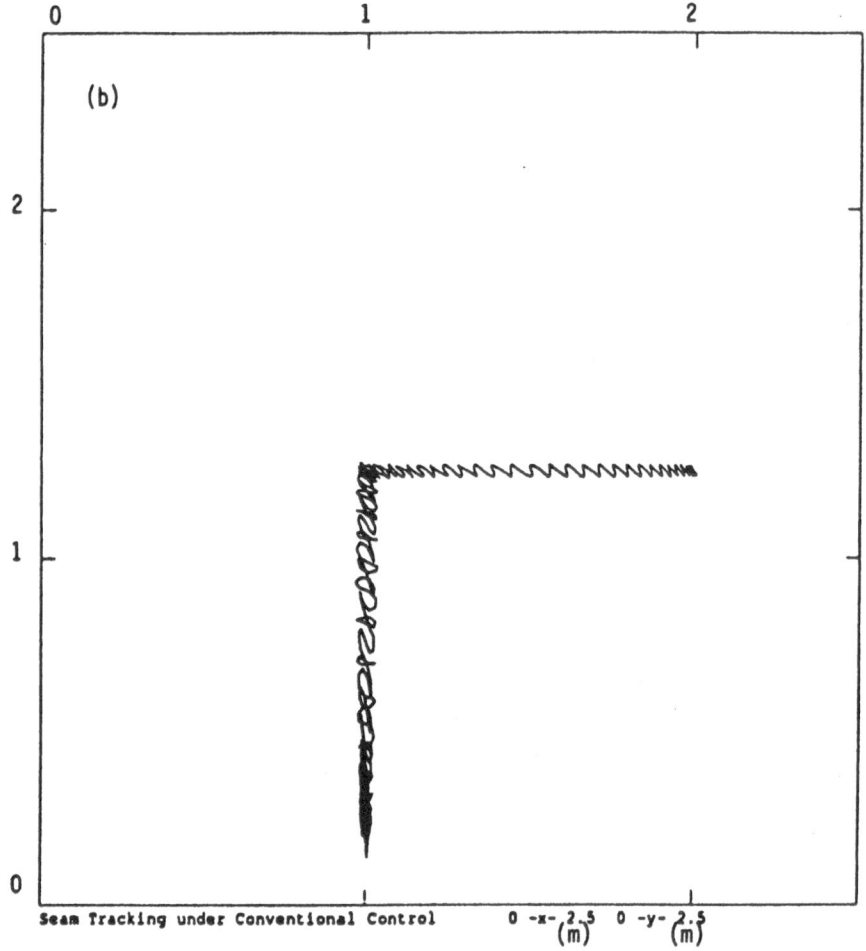

(b)

Seam Tracking under Conventional Control   0 -x- 2.5   0 -y- 2.5
                                               (m)   (m)

Figure 6.8(b)

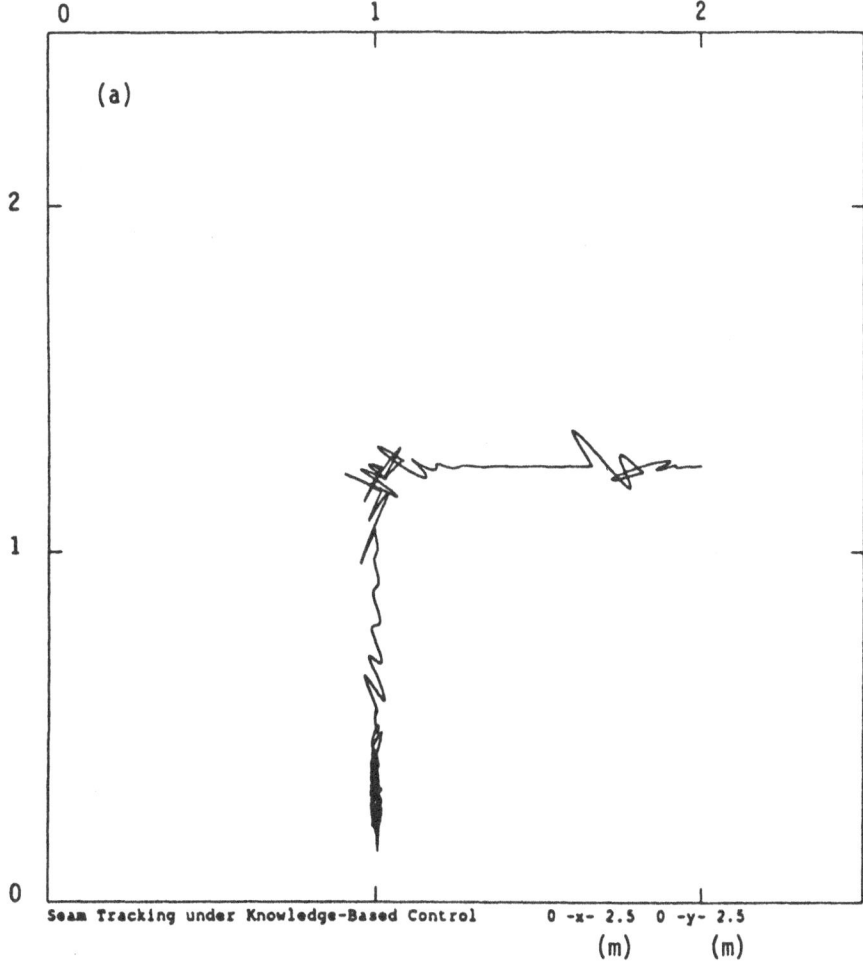

(a)

Seam Tracking under Knowledge-Based Control   0 -x- 2.5   0 -y- 2.5
                                                      (m)   (m)

Figure 6.9   Seam Tracking Subjected to
              Independent Joint Disturbances
              (a) Under Knowledge-Based Control
              (b) Under Conventional Control

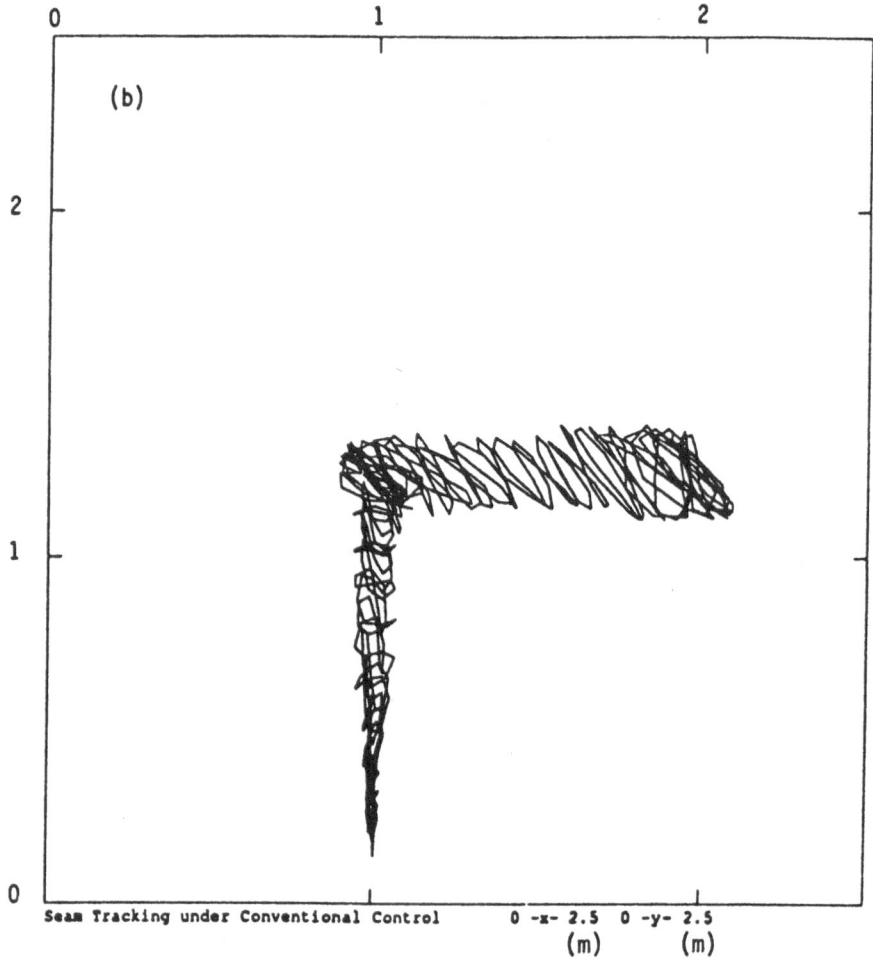

(b)

Seam Tracking under Conventional Control　　0 -x- 2.5　0 -y- 2.5
　　　　　　　　　　　　　　　　　　　　　　　　　　　(m)　　　(m)

Figure 6.9(b)

In the next experiment, the injected acceleration disturbance sequence was modified in the following way. The two joints were disturbed simultaneously at the time points 0.8 sec., 1.6 sec., 2.4 sec., and 3.4 sec. using the pairs (Joint 1, Joint 2) of acceleration pulses (0.05, 0.1), (-0.05, -0.1), (0.1, 0.05), and (-0.1, -0.05) given in rad/s^2. This simulation may represent, for example, two different pick-and-place operations within the same task. The results of this experiment are given in Figure 6.10. The response under conventional control appears to deteriorate through each successive disturbance, while the knowledge-based controller makes a good effort to improve the robot response after each disturbance.

Finally, a faster direction-change operation was simulated, under the same sequence of disturbances as in the previous simulation. Specifically, the end effector was uniformly accelerated during the first 1.5 sec. and rapidly decelerated to rest during the next 0.5 sec. at the corner. Then the end effector is is turned through 90° and rapidly accelerated to its full speed in 0.5 sec., and subsequently brought to rest at the end of the path during the final 1.5 sec. The results of this experiment are shown in Figure 6.11. Note the presence of a high frequency trajectory error in the neighbourhood of the corner, due to rapid deceleration and acceleration in that region. The knowledge-based controller seems to cope better with these high-frequency excitations.

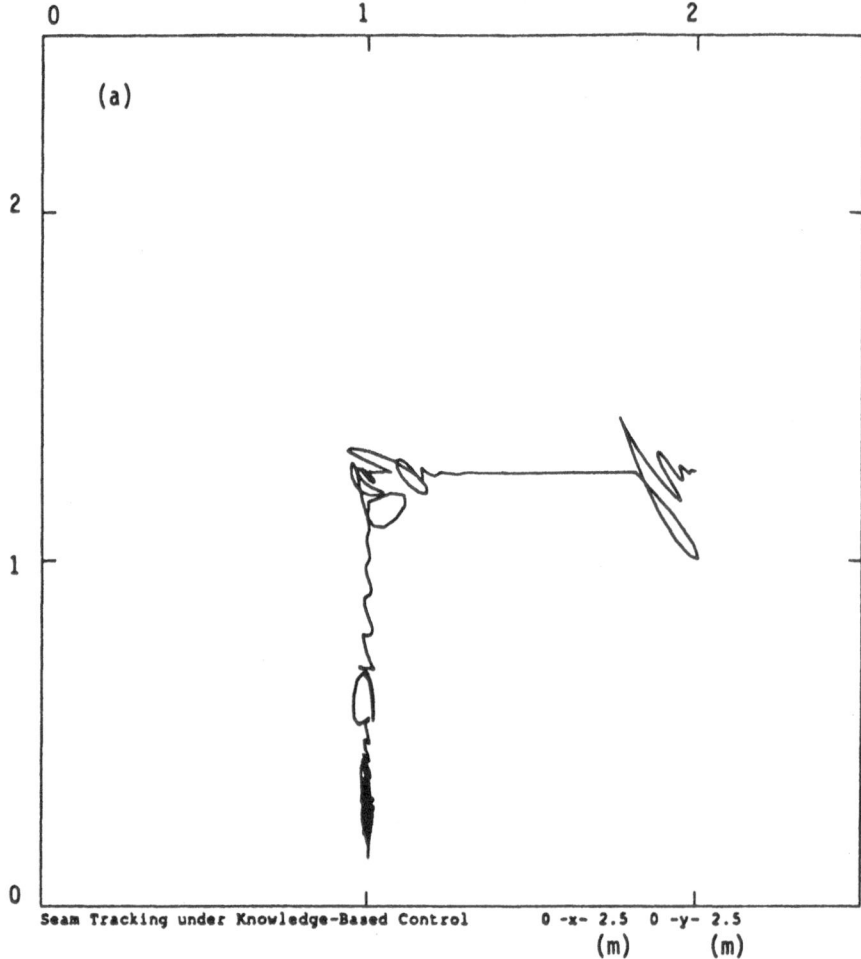

Figure 6.10   A Task with Two Pick-and-Place Operations
                  (a) Under Knowledge-Based Control
                  (b) Under Conventional Control

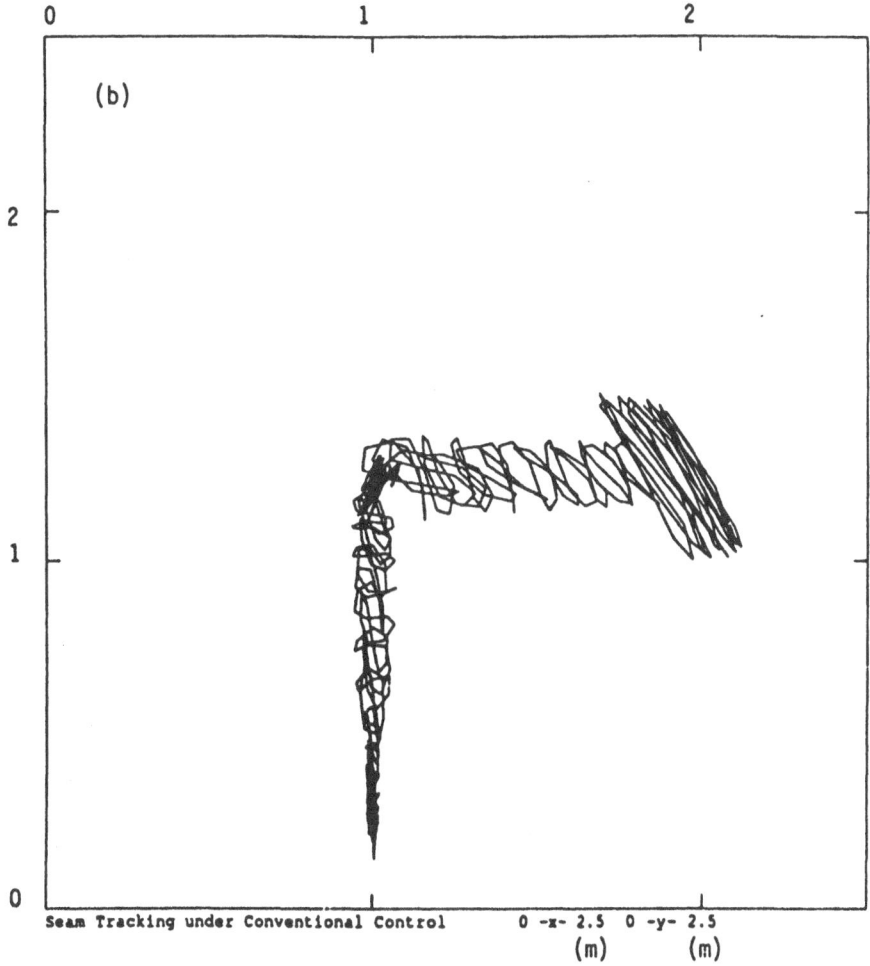

(b)

2

1

0

Seam Tracking under Conventional Control   0 -x- 2.5   0 -y- 2.5
                                                   (m)   (m)

Figure 6.10(b)

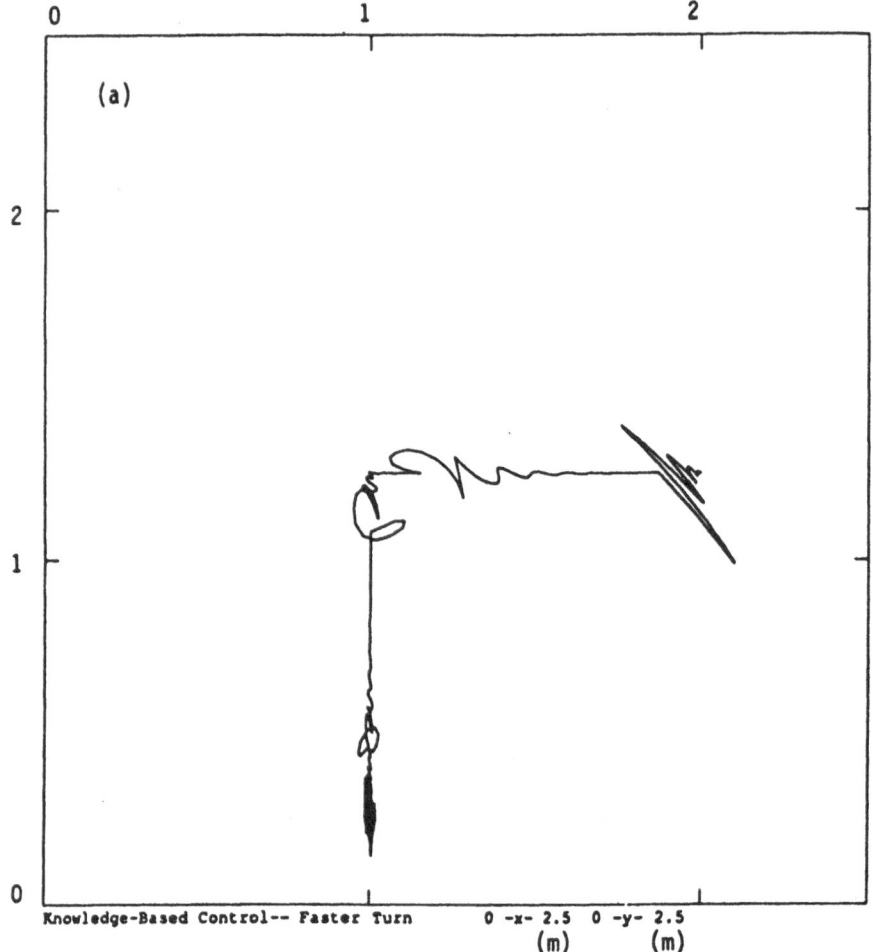

Knowledge-Based Control-- Faster Turn   0 -x- 2.5   0 -y- 2.5
                                             (m)   (m)

Figure 6.11   Pick-and-Place Operations with Fast Negotiation
               of a Corner
               (a) Under Knowledge-Based Control
               (b) Under Conventional Control

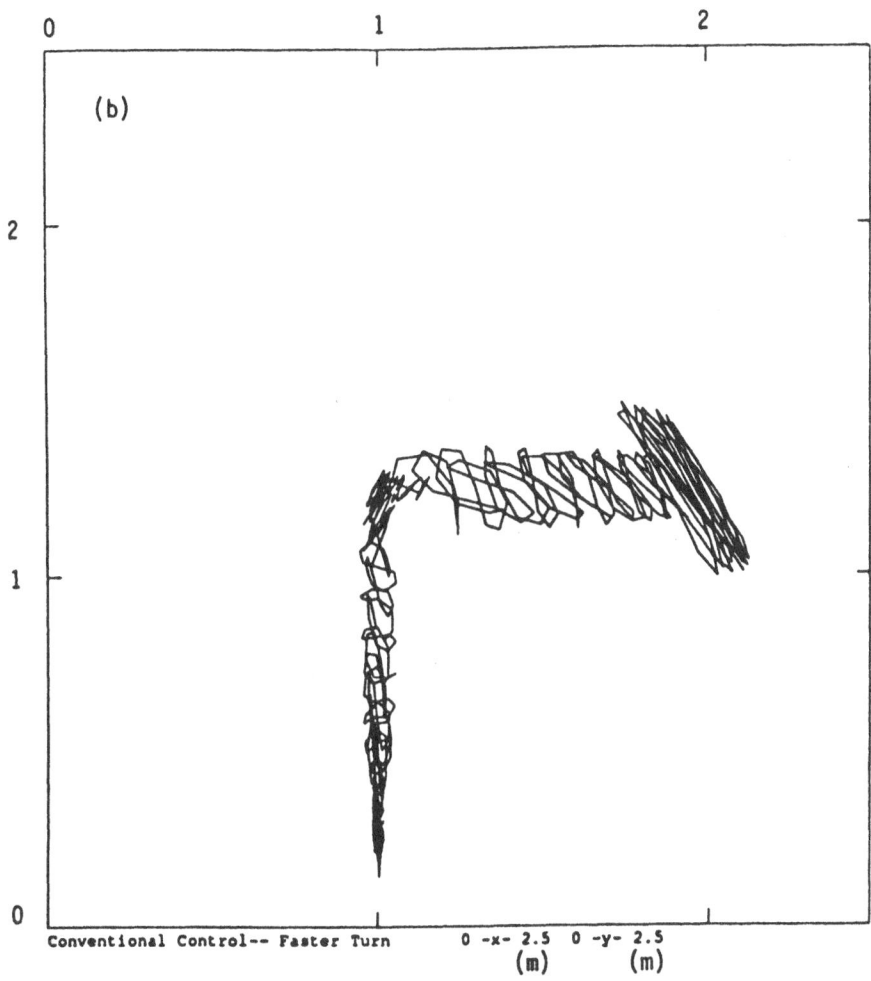

Figure 6.11(b)

6.5 Performance and Limitations

The simulation experiments presented in this chapter clearly show that the proposed knowledge-based controller is generally superior to a conventional controller which employs a low-level hard algorithm. But it remains to answer the question: at what added cost and complexity is this improvement in performance achieved? We shall now address some of the relevant issues of the performance and limitations of the proposed controller, particularly in the context of the application which was developed in Chapter 5 and the results which have been presented in this chapter. More fundamental issues, pertaining to the general control structure described in Chapter 4, will be discussed in Chapter 7.

The knowledge-based controller was found to quickly identify important trends or attributes of the robot response. The results show that oscillations, slow error convergence, steadily diverging unstable behaviour, and steady offsets in the joint responses have all been identified by the joint servo experts, as expected. Also, the servo tuning decisions made by the fuzzy controller were found to be satisfactory.

Note that the rule bases used in the servo experts and the linguistic rules in the fuzzy knowledge base are the key to the performance of the overall controller. Once these rules are established, not much effort was needed to "tune" the knowledge-based controller to achieve a satisfactory performance. In fact, the very first set of trial parameter values was able to generate good results, in the first series of simulations. This shows that much expertise is not required in selecting parameter values in the beginning, and that reasonable guesses seem to work well. For example, the same fuzzy decision table was used in all the simulations, and it contained the very first trial values used in its development. Furthermore, the first trial set of servo specifications which were used resulted in an acceptable performance. Of course, servo specifications have to be modified depending on the required stringency and tolerances of control, and on the physical nature of the process. Another encouraging observation was that the knowledge-based controller was quite robust and not very sensitive to the initial values of the servo parameters.

The fuzzy tuner need not operate at the bandwidth of the servo controller. Simulations showed that the bandwidth of the knowledge-based controller could be lower by at least an order of magnitude. In fact, some of the experiments described in the present chapter were run with a knowledge-based control bandwidth which was smaller than the servo control bandwidth by a factor of 16. But, generally, an improved performance was obtained by increasing the bandwidth of the knowledge-based controller. The control bandwidth was not a limitation in the present application because the rule base used was not extensive and consequently the rule processing time was quite small. Furthermore, the linguistic fuzzy rules were preprocessed to generate a set of decision tables which could be read very fast, during real-time operation. It is anticipated, however, that if a very low bandwidth is used, the knowledge-based controller would become more sensitive to the initial values of the servo parameters.

Test inputs of step, ramp, and sine were used to evaluate the performance of the individual joints. Seam tracking experiments were used to study the behaviour of the overall robot. In seam tracking experiments, the trajectory of the end effector was specified. The joints of the robot were expected to perform in a coordinated manner, to produce the specified end-effector trajectory. In addition to coupling of the servo experts introduced in this manner, another form of dynamic coupling was introduced through joint disturbances. As discussed in Chapter 4, coupling errors in a robot, under nonlinear feedback control, can be expressed as a vector of acceleration disturbances at the joints of the robot. Acceleration disturbances were therefore injected to the joints to simulate the effects of various tasks such as pick-and-place operations, as well as of coupling errors. Some of these results show that the knowledge-based controller somewhat overtunes the servos on the occurence of the first acceleration disturbance so that subsequent acceleration disturbances would be rapidly compensated for. Actually, a better performance could be obtained by modifying the rule base to anticipate such disturbances (and coupling errors) and to take corrective actions accordingly. Such a scheme was not implemented in the present study.

There are two primary limitations of the present simulation studies. First, the nonlinear feedback control algorithm was not implemented. Implementation and evaluation of the nonlinear feedback control scheme has to be done separately, so that any problems due to that control scheme would not overshadow the performance of the knowledge-based controller. Since a recursive algorithm has been developed in the present work, the rest of the implementation of the nonlinear feedback control scheme would require only a routine extension. Evaluation of the nonlinear feedback control scheme was not an objective of the present work. Second, an actual physical implementation and testing of the application was not undertaken. Of course, there are several additional issues which have to be addressed in a physical implementation. In particular, response sensing, interfacing, signal scaling, and selection of parameter values have to be done properly, for an available robot. That would require additional resources, but would be a typical project in prototyping any controller. Facilities are available from the suppliers of the MUSE AI Toolkit, for direct loading of the knowledge-based controller (servo experts and the fuzzy tuner) from the SUN workstation into a target machine (typically, a single-board computer) for subsequent interfacing with a process.

7. CONCLUSIONS

7.1 Introduction

This concluding chapter outlines the relevance and significance of the work which has been described in the monograph. In particular, it highlights the merits of the proposed knowledge-based control structure, and summarises the contributions made. Those aspects that were not addressed in detail are noted, and suggestions for future work are given.

7.2 Significance

The problem of integrating knowledge-based control into the conventional hard algorithms of direct digital control has been addressed in this monograph. Although an application of the proposed approach was demonstrated for a robotic manipulator, it is anticipated that the general control structure described here would be effective in many other types of moderate to high bandwidth processes possessing nonlinear and dynamically coupled characteristics. The motivation for the present research came from the observation that knowledge-based control is generally not effective as a direct substitute for conventional methods of standard digital control, in moderate to high bandwidth processes, and from the fact that linear servo control is normally unsatisfactory in nonlinear and coupled processes of this type. This, combined with the demonstrated fact that human experts can perform controller tuning tasks effectively, provided the impetus for the approach used.

The proposed knowledge-based controller has three hierarchical levels. At the lowest level there are conventional servo controllers which are closed around a high-speed nonlinear feedback controller. The second level is the first of two knowledge-based levels in the control structure. This level is somewhat like a knowledge-based preprocessor for measured process information. A servo expert is assigned to each degree of freedom uncoupled by the nonlinear feedback controller. It monitors the response of the degree of freedom and, on the basis of a set of performance specifications, makes inferences as to trends and characteristics of the response, in terms of a set of useful attributes. These inferences (attribute values) are passed on to the top level of the hierarchy, the fuzzy control level. A fuzzy tuner at the top level uses the inferences made by the servo experts to make tuning decisions. Particularly, the fuzzy tuner will modify the values of the servo parameters. In a more general context, the fuzzy tuner may be assigned many other tasks. It can update parameter values of the nonlinear feedback controller and even modify the initial performance specifications provided by the user (in Level 2). It may perform self-organisation and conflict-resolution tasks as well, perhaps employing additional information including external sensory data and expectational knowledge. For example, the rule bases of the servo experts and the linguistic rules of the fuzzy controller itself, could be modified. Furthermore, priorities could be assigned to various degrees of freedom (servos) on the basis of, say, measured outputs other than the responses of the degrees of freedom.

Needless to say it is not feasible to assign a human servo expert to every process that demands a servo expert. In low-bandwidth process control practice it is customary to use general guidelines provided by the supplier of a servo controller, for tuning the controller. Typically, tuning is done manually by a trained person (not necessarily an expert). In high-bandwidth processes, particularly those exhibiting nonlinear and coupled characteristics, manual tuning of servos might not provide the required performance accuracy. Hard algorithms have been used as adaptive controllers to perform tuning at high speed, in such situations. But a major shortcoming of hard algorithms is that they are not appropriate for representing the "soft" and "fuzzy" knowledge of a human expert. For example, it would not be possible to directly translate a tuning knowledge base available as a set of linguistic statements, into a hard algorithm. Use of a knowledge-based approach, in particular a fuzzy system, for high-level tuning, would be attractive for these reasons.

Other attempts have been made to replace the hard algorithms of direct digital control by soft knowledge-based methods. Serious drawbacks of such approaches have been recognised in selecting the control structure proposed here. Notably, the control bandwidth can deteriorate significantly by including a soft controller at the servo level, and in real-time direct digital control it is generally not possible to arbitrarily choose a control bandwidth, or to use arbitrarily variable sampling rates. Furthermore, by including a knowledge-based controller into the servo loop, "soft errors" due to vagueness will be directly introduced into drive signals. Also, since any knowledge-based controller requires a finite learning period, and since they are relatively insensitive to initial values, they are more appropriate as tuning controllers. In addition, for a high-bandwidth process, it is not practical to gain experience by manually sequencing the drive signals and manually observing the resulting responses. The proposed control structure is attractive in all these respects.

Flexibility of the control structure, and the relative ease of further development and modification are further advantages. Typically, one servo expert is developed and tested using suitable inputs, and then it is duplicated for the remaining degrees of freedom of the process. Since all degrees of freedom might not possess similar characteristics, each servo expert has to be examined separately and modified where necessary. Subsequently, parameter values have to be assigned for the servo experts by taking into account the characteristics of individual degrees of freedom (e.g., natural frequencies) and the performance requirements (e.g., sensor error tolerances). The development process can be significantly expedited in this manner, particularly for complex and high-order systems. Flexibility of the control structure stems mainly from the ease with which old knowledge could be modified and new knowledge could be added. Since knowledge is present as a set of rules, one is able to simply add or delete appropriate rules. Furthermore, there are built-in mechanisms such as conflict resolution schemes to warn about incompatible, redundant, or erroneous rules within the rule base. For example, if it was found that a particular rule was not fired after prolonged operation of the controller, one should reexamine that particular rule for its validity. As another example, if the upper bound of a membership function estimate, for a control action variable, is not close to unity, then this is an indication that the fuzzy rule base is incomplete.

One could argue that, once an effective nonlinear feedback controller has been implemented on a process, a satisfactory performance might be achieved by properly choosing the parameter values for the servos and setting them at the beginning of each task, thereby eliminating the need for a knowledge-based tuner. This argument does not hold generally, and even when it holds it is circular in nature. Specifically, some knowledge is needed to choose "proper" values for the servo parameters. The knowledge-based controller can start with a more or less arbitrary set of values and still give good performance. Furthermore, if the servo settings are overspecified, the conventional approach would give good performance but at the cost of an excessive control effort. The knowledge-based controller will adjust such overspecified conditions as well. Also, generally, it is not possible to pick one set of servo parameters which is optimal over a wide range of operating conditions.

In the work described here, the development of the proposed control structure was carried out in the context of an application to robotic manipulators. Robots are nonlinear and coupled processes of typical bandwidth in the neighbourhood of 100 Hz. Servo control is widely used in the low-level control of industrial robots and there has been a significant research interest in augmenting servo controllers with hard algorithms of the adaptive control and nonlinear feedback control types. Consequently, robots are ideal candidates for exploiting the benefits of the proposed control structure.

7.3 Contributions

The proposed knowledge-based control structure represents the main contribution of the work described here. Theory, concepts, and procedures for the development of each level in the control structure have been presented. An application of the knowledge-based controller has been described with reference to a robotic manipulator.

At the knowledge-based level, the control hierarchy is divided into two levels. The first level is an intelligent preprocessor which makes inferences on the basis of a set of servo specifications. The top level is a fuzzy tuner which, by its very nature, can represent a body of tuning knowledge expressed in terms of linguistic statements. This division of the knowledge-based tasks into two levels is considered to be an important contribution of this study.

Furthermore, soft knowledge-based tasks are recognised as low-bandwidth tasks. Tuning tasks are exclusively placed in that category. High-bandwidth direct digital control tasks are restricted to the lowest level. A nonlinear feedback controller has been employed to decouple and linearise the process, so that the use of linear servo controllers would be justified.

A demonstration of the steps involved in developing a control structure of the proposed type has been provided using a robotic manipulator example. A commercial AI toolkit has been used to develop and implement low-level knowledge-based tasks. Fuzzy control techniques have been used to develop decision tables for servo tuning.

A recursive algorithm has been developed for implementation of the nonlinear feedback control method in robotic manipulators. Efficient algorithms of this nature would be essential at the direct-digital-control level where a high control bandwidth is needed. A set of steps has been developed which could be used in a recursive algorithm, for compensating for backlash effects in robotic manipulators.

It is considered that a main obstacle to the widespread use of fuzzy control is difficulty in understanding the underlying mathematical concepts. To help alleviate this difficulty to some extent, simplified geometric representations have been provided for the fundamental fuzzy concepts which are used here. Notably, the compositional rule of inference has been interpreted as a rule-matching procedure. Several examples have been given to illustrate various fuzzy concepts.

7.4 Future Developments

Future developments in knowledge-based control should aim at incorporating the well established merits of hard control in improving the overall effectiveness of a control system. The simple approach of replacing hard control by direct knowledge-based control in order to make the controller "intelligent", cannot be justified in general. The promise of knowledge-based control lies primarily in its ability to use knowledge available in a nonnumeric form. Generally, reasoning and associated search procedures in knowledge-based control can be quite slow. Control structures of the form proposed here are important from this point of view because knowledge-based tasks are restricted to slow control zones in these structures. But, process performance can be improved by increasing the speed of the relevant knowledge-based tasks. Hence, it is important to explore possibilities of reducing the real-time processing overhead of a knowledge-based controller. The decision table approach used in the top level of the proposed control structure, which diverts most of the processing requirements to off line, is one such possibility. These real-time control issues have to be carefully studied. When the bandwidth of the knowledge-based structural levels are brought close to that of the hard-control level, there exists the danger of undesirable interactions among control zones in the structure. This is another area which needs further investigation.

The control structure proposed in this monograph could be further enhanced and evaluated by considering more extensive applications. For example, an implementation of self-organising capabilities, such as on-line modification of the rule bases, could be undertaken. Enhancement of the knowledge base to detect the level of the coupling errors that are present in a typical case of an imperfect nonlinear feedback controller would be useful.

Use of expectational knowlege to improve the performance of the knowledge-based controller is another aspect which deserves exploration. For example, with available knowledge of the nature of a task, it is possible to anticipate some types of external disturbances as well as possible abnormal behaviour. This would be the case in a pick-and-place operation of a robot where impact type disturbances would arise, or in an assembly operation where parts jamming could occur. The rule bases could be appropriately modified to handle such anticipated conditions.

The recursive algorithm which has been developed here, for the nonlinear feedback control of robotic manipulators, has to be implemented and its performance has to be evaluated using simulation experiments. Accuracy considerations, possible numerical problems, and computational time issues related to nonlinear feedback control are all topics which deserve further investigation. Subsequently, the application of the proposed control structure could be extended to include the nonlinear feedback control algorithm.

The natural next step of the present study is the physical implementation of an application of the proposed control structure, to demonstrate the effectiveness of this control approach in practice. Since an application, developed first on a host machine, could be directly loaded into a target machine (typically a single-board computer) for interfacing with a physical process, only the routine problems of prototyping a controller are anticipated in the physical implementation of the proposed system.

REFERENCES

1. Asada, H., Kanade, T., and Takeyama, I. "Control of a Direct-Drive Arm", *ASME Journal of Dynamic Systems, Measurement, and Control*, 105(3), 136-142, 1983.

2. Asada, H., and Slotine, J.J.E., *Robot Analysis and Control*, Wiley, New York, 1986.

3. Denavit, J., and Hartenberg, R.S., "A Kinematic Notation for Lower-Pair Mechanisms Based on Matrices", *ASME Journal of Applied Mechanics*, 22, 215-221, June 1955.

4. de Silva, C.W., "A Motion Control Scheme for Robotic Manipulators", *Proc. 1984 Canadian CAD/CAM and Robotics Conf.*, 13, 131-7, Toronto, June 1984.

5. _____, "Motion Sensors for Industrial Robots", *Mechanical Engineering*, ASME, 107(6), 40-51, June 1985.

6. _____, " Advanced Techniques for Robotic Manipulator Control", *Proc. 1986 Int. Congress on Tech. and Tech. Exchange*, 148-153, Pittsburgh, October 1986.

7. _____, *Control Sensors and Actuators* , Prentice-Hall, Inc., Englewood Cliffs, New Jersey, 1988.

8. de Silva, C.W. and Van Winssen, J.C., "A Recursive Algorithm for Least Squares Trajectory Control of Robotic Manipulators", *Proc. American Control Conf.*, 3, 1722-1727, Seattle, June 1986.

9. de Silva, C.W., Price, T.E., and Kanade, T., " Torque Sensor for Direct Drive Manipulators", *ASME Journal of Engineering for Industry*, 109(2), 122-127, May 1987.

10. de Silva, C.W. and Van Winssen, J.C., "Least Squares Adaptive Control for Trajectory Following Robots", *ASME Journal of Dynamic Systems, Measurement, and Control*, 109(2), 104-110, June 1987.

11. de Silva, C.W., Chung, C.L., and Lawrence, C., "Base Reaction Optimization of Robotic Manipulators for Space Applications", *Proc. Int. Symposium on Robots,* Sydney, November 1988 (In press).

12. Dubois, D. and Prade, H., *Fuzzy Sets and Systems*, Academic Press, Orlando, 1980.

13. Dubowski, S. and Des Forges, D.T., "The Application of Model-Referenced Adaptive Control to Robotic Manipulators", *ASME Journal of*

Dynamic Systems, Measurement, and Control, 101, 193-200, September 1979.

14. Francis, J.C. and Leitch, R.R., "ARTIFACT: A Real-time Shell for Intelligent Feedback Control", *Research and Development in Expert Systems*, Bramer, M.A. (ed.), Cambridge University Press, 1985.

15. Freedy, A., "Learning Control in Remote Manipulators and Robot Systems", *Learning Systems*, 53-69, American Control Council, New York, 1973.

16. Fu, K.S., Gonzalez, R.C., and Lee, C.S.G., *Robotics*, McGraw-Hill, New York, 1987.

17. Goff, K.W., "Artificial Intelligence in Process Control", *Mechanical Engineering*, ASME, 107(10), 53-57, October 1985.

18. Gupta, M.M. and Sanchez, E. (eds.), *Fuzzy Information and Decision Processes*, North-Holland, Amsterdam, 1982.

19. Hemami, H. and Camana, P.C., "Nonlinear Feedback in Simple Locomotion Systems", *IEEE Trans. on Automatic Control*, AC-21(6), 855-860, December 1976.

20. Hewit, J.R. and Burdess, J.S., "Fast Dynamic Decoupled Control for Robotics using Active Force Control", *Mechanism and Machine Theory*, 16(5), 535-542, 1981.

21. Hirota, K., Arari, Y., and Pedrycz, W., "Robot Control Based on Membership and Vagueness", *Approximate Reasoning in Expert Systems*, Gupta, M.M., et. al. (eds.), 621-635, Elsevier, Amsterdam, 1985.

22. Hollerbach, J.M., "A Recursive Formulation of Lagrangian Manipulator Dynamics", *Proc. Joint Automatic Control Conf.*, TP10-B, San Francisco, 1980.

23. Holmblad, I.P. and Ostergaard, J.J., "Fuzzy Logic Control: Operator Experience Applied in Automatic Process Control", *FLS Review*, Copenhagen, 1981.

24. Horn, B.K.P. and Raibert, M.H., "Manipulator Control Using the Configuration Space Method", *The Industrial Robot*, 5(2), 69-73, June 1978.

25. Isik, C. and Mystel, A., "Decision Making at a Level of Hierarchical Control for Unmanned Robots", *Proc. 1986 IEEE Int. Conf. on Robotics and Automation*, 1772-1778, IEEE Computer Society, Los Angeles, 1986.

26. Kahn, M.E., "The Near-Minimum-Time Control of Open-Loop Articulated Chains", *A.I. Memo 177*, Stanford AI Laboratory, California, December 1969.

27. Khosla, P., *Real-Time Control and Identification of Direct-Drive Manipulators*, Ph.D. Thesis, Dept. Elec. and Comp. Engineering, Carnegie Mellon University, Pittsburgh, August 1986.

28. Kornblugh, R.D., *An Experimental Evaluation of Robotic Manipulator Dynamic Performance under Model Referenced Adaptive Control*, S.M. Thesis, Dept. Mech. Engineering, Massachusetts Institute of Technology, Cambridge, September 1984.

29. Luh, J.Y.S., Walker, M.W., and Paul, R.P.C., "On-line Computation Scheme for Mechanical Manipulators", *ASME Journal of Dynamic Systems, Measurement, and Control*, 102, 69-76, June 1980.

30. Lynch, P.M., "Minimum-time Sequential Axis Operation of a Cylindrical Two-axis Manipulator", *Proc. Joint Automatic Control Conf.*, 1(WP-2A), Charlottesville, 1981.

31. Mason, M.T. and Salisbury, J.K., *Robot Hands and the Mechanics of Manipulation*, The MIT Press, Cambridge, 1985.

32. Mamdani, E.H., "Application of Fuzzy Logic to Approximate Reasoning using Linguistic Synthesis", *IEEE Trans. on Computers*, C-26(12), 1182-1191, December 1977.

33. Mamdani, E.H. and Gaines, B.R. (eds.), *Fuzzy Reasoning and Its Applications*, Academic Press, London, 1981.

34. *MUSE Support System Manual Set*, Cambridge Consultants Ltd., Cambridge, U.K., March 1987.

35. Paul, R., "The Mathematics of a Computer Controlled Manipulator", *Proc. Joint Automatic Control Conf.*, 1, 124-131, 1977.

36. _____, *Robot Manipulators*, The MIT Press, Cambridge, 1981.

37. Procyk, T.J. and Mamdani, E.H., "A Linguistic Self-Organizing Controller", *Automatica*, 15, 15-30, 1979.

38. *PROTUNER 1100 Instruction Manual Set*, Techmation Inc., Tempe, 1984.

39. Raibert, M.H. and Craig, J.J., "Hybrid Position/Force Control of Manipulators", *ASME Journal of Dynamic Systems, Measurement, and Control*, 102, 126-133, June 1981.

40. Scharf, E.M. and Mandic, N.J., "The Application of a Fuzzy Controller to the Control of a Multi-degree-of-freedom Robot Arm", *Industrial Applications of Fuzzy Control,* Sugeno, M. (ed.), North-Holland, Amsterdam, 41-61, 1985.

41. Shibly, H.A., *Performance Evaluation and Efficient Control of Trajectory Following Robots with Friction and Backlash,* Ph.D. Thesis, Dept. Mech. Engineering, Carnegie Mellon University, Pittsburgh, February 1988.

42. Slotine, J.J.E., "The Robust Control of Robot Manipulators", *The Int. Journal of Robotics Research,* 4(2), 49-64, Summer 1985.

43. Staugaard, A.C., *Robotics and AI,* Prentice Hall, Englewood Cliffs, 1987.

44. Sugeno, M. (ed.), *Industrial Applications of Fuzzy Control,* North-Holland, Amsterdam, 1985.

45. Tokumaru, H. and Iwai, Z., "Noninteracting Control of Nonlinear Multivariable Systems", *Int. Journal of Control,* 16(5), 945-958, 1972.

46. Tong, R.M., "A Control Engineering Review of Fuzzy Systems", *Automatica,* 13, 559-569, 1977.

47. Uicker, J.J., *On the Dynamic Analysis of Spatial Linkages using 4x4 Matrices*, Ph.D. Thesis, Northwestern University, Evanston, August 1965.

48. Van Amerongen, J., Van Nauta Lemke, H., and Van der Veen, J.C.T., "An Autopilot for Ships Designed with Fuzzy Sets", *Proc. Fifth IFAC/IFIP Int. Conf. on Digital Computer Appl. Process Control*, 1977.

49. Van Brussel, K. and Vastmans, L., "A Compensation Method for the Dynamic Control of Robots", *Proc. Conf. Robotics Research*, MS 84-487, Bethlehem, PA, August 1984.

50. Whitney, D.E., "Resolved Motion Rate Control of Manipulators and Human Prosthesis", *IEEE Trans. on Man-Machine Systems*, MMS-10(2), 47-53, June 1969.

51. _____, "Historical Perspective and State of the Art in Robot Force Control", *The Int. Journal of Robotics Research*, 6(1), 3-14, Spring 1987.

52. Wu, C.H. and Paul, R.P., "Resolved Motion Force Control of Robot Manipulators", *IEEE Trans. Systems, Man, and Cybernetics*, SMC-12(3), 266-275, May 1982.

53. Yoshikawa, T., "Analysis and Control of Robot Manipulators with Redundancy", *Robotics Research,* Brady, M. and Paul, R. (eds.), The MIT Press, Cambridge, 1984.

54. Young, K.K.D., "Controller Design for a Manipulator using Theory of Variable Structure Systems", *IEEE Trans. System, Man, and Cybernetics,* SMC-8(2), 101-109, February 1978.

55. Zadeh, L.A., "From Circuit Theory to System Theory", *Proc. Institute of Radio Engineers,* 50, 856-865, 1962.

56. Zimmerman, H.J., Zadeh, L.A., and Gaines, B.R. (eds.), *Studies in the Management Sciences,* North-Holland, Amsterdam, 1984.

Appendix 1. Program Listing

Listings of the computer programs developed for the application described in Chapter 5 and Chapter 6, are given in this appendix. The programs included are the PopTalk code of the servo experts and fuzzy tuner, PopTalk interface programs, and the C program of the robot simulator.

```
/*****************************************************************/
/*                                                             */
/*          THE KNOWLEDGE-BASED CONTROL PROGRAMS               */
/*                                                             */
/*****************************************************************/


/********************************************/
/*                                          */
/*      The definition of the schemas       */
/*                                          */
/********************************************/

global 0 0 system 040886
~ object System_object_type ^System_object {
  system 100286 | FUBAR
~~declarations
    /* Forward declarations go in here */
~~internal_schemas
#1
~~schemas
~    collection generic_schema {
~       object generic_schema ^inference {
        desilva 120388 | Inferences of Joint Ruleset
~~        Comment
            Inferences made by the ruleset
            at each joint of robot regarding the error response.
~~       isa
          DB_obj
~~       slot name
          "noname"
~~       slot decay
          "pending"
~~       slot divergence
          "unknown"
~~       slot oscillation
          "unknown"
~~       slot offset
          "unknown"
~~       slot accuracy
          "unknown"
}
~       object generic_schema ^specs {
        desilva 120388 | Joint response specs.
~~        Comment
            Specifications on error decay,
            oscillation amplitude, and offset.
~~       isa
          DB_obj
~~       slot name
          "noname"
~~       slot lambda
          0.0
~~       slot ampl
          0.0
~~       slot ess
            0.0
}
~       object generic_schema ^error {
        desilva 120388 | Monitored error samples.
~~        Comment
            Sampling period and three
```

```
                    successive values of maximum error, minimum error,
                    and maximum
                    error over three successive periods.
~~       isa
                    DB_obj
~~       slot name
                    "noname"
~~       slot period
                       0.4
~~       slot e0
                    0.0
~~       slot e1
                    0.0
~~       slot e2
                    0.0
}
}
~~instances
~    collection   {
}
~~libraries
~    collection Library_File {
~        object Library_File /usr/local/Muse/poptalklib
                                  /sockets.pt {
         desilva 120688 | Pop source file
}
}
~~knowledge_sources
#2
~~notice_boards
~    collection NB {
~        object NB spec_nb1 {
         desilva 120388 | Joint 1 response specs.
~~        Comment
                    Specifications on error response
                    at joint 1 of robot.
~~        initial_entries
~            collection   {
~              object specs spec1 {
              desilva 120388 | Joint 1 response specs.
~~             Comment
                    Specifications on error response
                    at joint 1 of robot.
~~             lambda
                       1.0
~~             ampl
                       0.005
~~             ess
                       0.01
~~             name
                    "spec1"
}
}
~~        demons
~            collection demon {
}
}
~        object NB spec_nb2 {
         desilva 120388 | Joint 1 response specs.
~~        Comment
                    Specifications on error response
                    at joint 2 of robot.
~~        initial_entries
~            collection   {
~        object specs spec2 {
              desilva 120388 | Joint 1 response specs.
```

```
~~          Comment
                Specifications on error response
                at joint 2 of robot.
~~          lambda
                1.0
~~          ampl
                0.005
~~          ess
                0.01
~~          name
                "spec2"
}
}
~~      demons
~           collection demon {
}
}
}
~~user_init
      lambda;
      /*    if $stream_from.schemaof /== stream then */
            socket_link();
      /*    endif; */
      endlambda,
}


/****************************************/
/*                                      */
/*      The first knowledge source      */
/*                                      */
/****************************************/

global 3 2 desilva 120388
~ object KS servo_ks1 {
  desilva 120388 | Joint 1 knowledge source.
~~ Comment
        The knowledge source that makes
        inferences on the dynamic response at
        joint 1 of robot.
~~ priority
      high_priority
~~ libraries
~      collection Library_File {
~        object Library_File chan1.pt {
         desilva 120688 | Pop source file
}
}
~~ initial_entries
~      collection  {
~        object inference infr1 {
         desilva 200688 | inference made on error
~~          Comment
                Inference made by servo expert at Joint 1.
~~          decay
                "pending"
~~          divergence
                "unknown"
~~          oscillation
                "unknown"
~~          offset
                "unknown"
~~          accuracy
                "unknown"
~~          name
                "infr1"
```

```
}
}
~~  demons
~       collection demon {
}
~~  relations
~       collection  {
}
~~  reasoning_modules
~       collection  {
#4
}
}


/*************************************************/
/*                                               */
/*    The rule base of the first servo expert    */
/*    along with the fuzzy decision table        */
/*                                               */
/*************************************************/

global 4 3 desilva 120388
~ object FPRuleset pid_rules1 {
  desilva 120388 | Joint 1 response testing rules.
~~  Comment
       Rules that make inferences on
       the joint 1 response of robot.
~~  rules
~       collection FPR {
~          object FPR okay_rule1 {
           desilva 120388 | Accurate response.
~~           Comment
               Checks whether the joint 1
               response is accurate.
~~           Source
               if
                   there is an inference I
                       -name "infr1" and
                   there is a specs
                       -name "spec1",
                 -ess E and
                   there is an error ER
                       -name "error1",
                 -e0 A where (abs(A) =< E),
                 -e1 B where (abs(B) =< E),
                 -e2 C where (abs(C) =< E)
               then
                   do(
                   vars strm1 = {stream:};
                   open('data1', 1) -> strm1;
                   strm1: flushit;
                   strm1: putchar(' 0.00  0.00  0.00');
                   strm1: flushit;
                   strm1: closeit; )      and
                   (printf('Joint 1 response is accurate.
                           \n');) and
                   assert {inference I: -accuracy "okay"}
                   and
                   delete ER from KS
       }
~          object FPR oscillations_rule1 {
           desilva 120388 | Joint 1 oscillations.
~~           Comment
               Checks whether the error response
               at joint 1 has undesirable oscillations.
```

```
~~          Source
              if
                  there is an inference I
                      -name "infr1" and
                  there is a specs
                      -name "spec1",
                -ampl Am and
                  there is an error  ER
                      -name "error1",
                -e0 A,
                -e1 B,
                -e2 C where (((B < (A - 2.0 * Am)) and
                (C > (B + 2.0 * Am))) or
                ((B > (A + 2.0 * Am))
                and (C < (B - 2.0 * Am))))   then
                do(
                vars strm1 = {stream:};
                open('data1', 1) -> strm1;
                strm1: flushit;
                if abs(C - B) > 5.0 * Am
                then strm1: putchar('-0.50  0.00  0.50');
                else strm1: putchar('-0.30  0.00  0.30');
                endif;
                strm1: flushit;
                strm1: closeit; )  and
                  (printf('Joint 1 response has undesirable
                            oscillations. \n');) and
                  assert {inference I: -oscillation "present"}
                  and
                  delete ER from KS
}
~          object FPR nodecay_rule1 {
           desilva 120388 | Decay of Joint 1 error.
~~          Comment
             Checks whether the error response
             at joint 1 decays adequately.
~~          Source
              if
                  there is an inference I
                      -name "infr1" and
                  there is a specs
                      -name "spec1",
                -lambda L and
                  there is an error  ER
                      -name "error1",
                -period T,
                -e0 A,
                -e1 B,
                -e2 C where ((((A > 0.0) and (A > B)
                            and (B > C)) and
                ((B > (A * (1.0 - L * T))) or
                (C > (A * (1.0 - 2.0 * L * T))))) or
                (((A < 0.0) and (A < B) and (B < C)) and
                ((B < (A * (1.0 - L * T))) or
                (C < (A * (1.0 - 2.0 * L * T)))))))   then

                do(
                vars strm1 = {stream:};
                open('data1', 1) -> strm1;
                strm1: flushit;
                strm1: putchar(' 0.15  0.00  0.30');
                strm1: flushit;
                strm1: closeit; )  and
                  (printf('Joint 1 response does not decay
                            adequately. \n');) and
                  assert {inference I: -decay "unacceptable"}
```

```
                                  and
                                  delete ER from KS
}
~          object FPR divergence_rule1 {
           desilva 120388 | Divergence of Joint 1 error.
~~            Comment
                 Checks whether the error
                 response at joint 1 of robot diverges.
~~            Source
                 if
                         there is an inference I
                                  -name "infr1" and
                         there is an error  ER
                                  -name "error1",
                    -e0 A,
                    -e1 B
                    -e2 C where (((A >= 0.0) and (B > A)
                                        and (C > B)) or
                              ((A =< 0.0) and (B < A)
                                 and (C < B))) then

                              do(
                              vars strm1 = {stream:};
                              open('data1', 1) -> strm1;
                              strm1: flushit;
                              strm1: putchar('-0.30 -0.30  0.50');
                              strm1: closeit; )  and
                         (printf('Joint 1 error steadily diverges.
                                   \n');) and
                         assert {inference I: -divergence "present"}
                         and
                         delete ER from KS
}
~          object FPR offset_rule1 {
           desilva 120388 | Offset in joint 1 response.
~~            Comment
                 Checks whether there is a steady offset
                 in the response at joint 1 of robot.
~~            Source
                 if
                         there is an inference I
                                  -name "infr1" and
                         there is a specs
                                  -name "spec1",
                    -ampl Am,
                    -ess E and
                         there is an error  ER
                                  -name "error1",
                    -e0 A,
                    -e1 B,
                    -e2 C where (((A >= E) and (B >= E) and
                                        (C >= E) and
                              (A =< (E + Am)) and (B =< (E + Am)) and
                                    (C =< (E + Am))) or
                               ((A =< (-1.0 * E)) and (B =< (-1.0 * E))
                              and
                               (C =< (-1.0 * E)) and (A >= -1.0 *(E + Am))
                              and
                               (B >= -1.0*(E + Am)) and (C >= -1.0*(E + Am))))
                              then

                              do(
                              vars strm1 = {stream:};
                              open('data1', 1) -> strm1;
                              strm1: flushit;
                              strm1: putchar(' 0.15  0.60  0.00');
```

```
                            strm1: flushit;
                            strm1: closeit; ) and
                             (printf('Joint 1 response has an unacceptable
                                        offset. \n');) and
                            assert {inference I: -offset "present"}  and
                            delete ER from KS
      }
~             object FPR final_rule1 {
              desilva 130688 | If other rules are not fired
~~              Comment
                    Delete error object if rules are not fired
                    on Joint 1.
~~              Source
                    if ALONE
                          there is an error ER
                                -name "error1"  then

                          do(
                          vars strm1 = {stream:};
                          open('data1', 1) -> strm1;
                          strm1: flushit;
                          strm1: putchar(' 0.00  0.00  0.00');
                          strm1: flushit;
                          strm1: closeit; )       and
                           (printf('Rule search continued for Joint 1.
                                     \n');) and
                          delete ER from KS
      }
      }
~~    interests
              [spec_nb1]
      }



/********************************************/
/*                                          */
/*      The second knowledge source         */
/*                                          */
/********************************************/

global 5 2 desilva 230688
~ object KS servo_ks2 {
  desilva 120388 | Joint 1 knowledge source.
~~  Comment
        The knowledge source that makes
        inferences on the dynamic response at
        joint 2 of robot.
~~  priority
      high_priority
~~  libraries
~       collection Library_File {
~ object Library_File chan2.pt {
          desilva 120688 | Pop source file
  }
  }
~~  initial_entries
~       collection  {
~ object inference infr2 {
          desilva 200688 | inference made on error
~~          Comment
                Inference made by servo expert at Joint 2.
~~          decay
              "pending"
~~          divergence
              "unknown"
~~          oscillation
```

```
                "unknown"
~~          offset
                "unknown"
~~          accuracy
                "unknown"
~~          name
                "infr2"
}
}
~~  demons
~      collection demon {
}
~~  relations
~      collection  {
}
~~  reasoning_modules
~      collection  {
#6
}
}


/*****************************************************/
/*                                                   */
/*    The rule base of the second servo expert       */
/*    along with the fuzzy decision table            */
/*                                                   */
/*****************************************************/

global 6 5 desilva 230688
~ object FPRuleset pid_rules2 {
   desilva 120388 | Joint 1 response testing rules.
~~  Comment
        Rules that make inferences on
        the joint 2 response of robot.
~~  rules
~      collection FPR {
~ object FPR okay_rule2 {
        desilva 120388 | Accurate response.
~~        Comment
            Checks whether the joint 2
            response is accurate.
~~        Source
            if
                there is an inference I
                    -name "infr2" and
                there is a specs
                    -name "spec2",
              -ess E and
                there is an error ER
                    -name "error2",
              -e0 A where (abs(A) =< E),
              -e1 B where (abs(B) =< E),
              -e2 C where (abs(C) =< E)   then

                do(
                vars strm2 = {stream:};
                open('data2', 1) -> strm2;
                strm2: flushit;
                strm2: putchar(' 0.00  0.00  0.00');
                strm2: flushit;
                strm2: closeit; )       and
                (printf('Joint 2 response is accurate.
                        \n');) and
                assert {inference I: -accuracy "okay"}
                and
```

```
                delete ER from KS
}
~ object FPR oscillations_rule2 {
        desilva 120388 | Joint 1 oscillations.
~~          Comment
                Checks whether the error response
                at joint 2 has undesirable oscillations.
~~          Source
                if
                        there is an inference I
                            -name "infr2" and
                        there is a specs
                            -name "spec2",
                    -ampl Am and
                        there is an error  ER
                            -name "error2",
                    -e0 A,
                    -e1 B,
                    -e2 C where (((B < (A - 2.0 * Am)) and
                (C > (B + 2.0 * Am))) or ((B > (A + 2.0 * Am))
                and (C < (B - 2.0 * Am))))  then

                    do(
                    vars strm2 = {stream:};
                    open('data2', 1) -> strm2;
                    strm2: flushit;
                    if abs(C - B) > 5.0 * Am
                    then strm2: putchar('-0.50   0.00   0.50');
                    else strm2: putchar('-0.30   0.00   0.30');
                    endif;
                    strm2: flushit;
                    strm2: closeit; )  and
                      (printf('Joint 2 response has undesirable
                                oscillations. \n');) and
                      assert {inference I: -oscillation "present"}
                      and
                      delete ER from KS
}
~ object FPR nodecay_rule2 {
        desilva 120388 | Decay of Joint 1 error.
~~          Comment
                Checks whether the error response
                at joint 2 decays adequately.
~~          Source
                if
                        there is an inference I
                            -name "infr2" and
                        there is a specs
                            -name "spec2",
                    -lambda L and
                        there is an error  ER
                            -name "error2",
                    -period T,
                    -e0 A,
                    -e1 B,
                    -e2 C where ((((A > 0.0) and (A > B)
                                and (B > C)) and
                ((B > (A * (1.0 - L * T))) or
                 (C > (A * (1.0 - 2.0 * L * T)))))) or
                 (((A < 0.0) and (A < B) and (B < C)) and
                ((B < (A * (1.0 - L * T))) or
                 (C < (A * (1.0 - 2.0 * L * T))))))     then

                    do(
                    vars strm2 = {stream:};
                    open('data2', 1) -> strm2;
```

```
                    strm2: flushit;
                    strm2: putchar(' 0.15   0.00   0.30');
                    strm2: flushit;
                    strm2: closeit; )      and
                     (printf('Joint 2 response does not decay
                              adequately. \n');) and
                    assert {inference I: -decay "unacceptable"}
                    and
                    delete ER from KS
     }
~ object FPR divergence_rule2 {
         desilva 120388 | Divergence of Joint 1 error.
~~        Comment
               Checks whether the error
               response at joint 2 of robot diverges.
~~        Source
               if
                    there is an inference I
                        -name "infr2" and
                    there is an error  ER
                        -name "error2",
                 -e0 A,
                 -e1 B
                 -e2 C where (((A >= 0.0) and (B > A) and
                                 (C > B)) or
                        ((A =< 0.0) and (B < A) and (C < B)))
                        then

                        do(
                        vars strm2 = {stream:};
                        open('data2', 1) -> strm2;
                        strm2: flushit;
                        strm2: putchar('-0.30 -0.30   0.50');
                        strm2: flushit;
                        strm2: closeit; )      and
                     (printf('Joint 2 error steadily diverges.
                              \n');) and
                    assert {inference I: -divergence "present"}
                    and
                    delete ER from KS
     }
~ object FPR offset_rule2 {
         desilva 120388 | Offset in joint 1 response.
~~        Comment
               Checks whether there is a steady offset
               in the response at joint 2 of robot.
~~        Source
               if
                    there is an inference I
                        -name "infr2" and
                    there is a specs
                        -name "spec2",
                 -ampl Am,
                 -ess E and
                    there is an error  ER
                        -name "error2",
                 -e0 A,
                 -e1 B,
                 -e2 C where (((A >= E) and (B >= E) and
                                 (C >= E) and
                    (A =< (E + Am)) and (B =< (E + Am)) and
                    (C =< (E + Am))) or
                     ((A =< (-1.0 * E)) and (B =< (-1.0 * E)) and
                     (C =< (-1.0 * E)) and (A >= -1.0 *(E + Am))
                     and
                     (B >= -1.0*(E + Am)) and
```

```
                        (C >= -1.0*(E + Am))))   then

                        do(
                        vars strm2 = {stream:};
                        open('data2', 1) -> strm2;
                        strm2: flushit;
                        strm2: putchar(' 0.15  0.60  0.00');
                        strm2: flushit;
                        strm2: closeit; )   and
                         (printf('Joint 2 response has an unacceptable
                                   offset. \n');) and
                         assert {inference I: -offset "present"}   and
                         delete ER from KS
}
~ object FPR final_rule2 {
        desilva 130688 | If other rules are not fired
~~         Comment
               Delete error object if rules for Joint 2 are
               not fired.
~~         Source
               if ALONE
                       there is an error ER
                           -name "error2"      then

                       do(
                       vars strm2 = {stream:};
                       open('data2', 1) -> strm2;
                       strm2: flushit;
                       strm2: putchar(' 0.00  0.00  0.00');
                       strm2: flushit;
                       strm2: closeit; )       and
                       (printf('Rule search continued for Joint 2.
                                 \n');) and
                       delete ER from KS
}
}
~~   interests
        [spec_nb2]
}
```

```
/*********************************/
/*                             */
/*      THE INTERFACE PROGRAMS    */
/*                             */
/*********************************/

/*****************************************************/
/*                                                 */
/*      The interface from the robot simulator      */
/*      to the first servo expert                   */
/*                                                 */
/*****************************************************/

vars error1 = {error:
                 -name "error1",
                 -period 0.4};
vars servo_ks1 ={KS:};

vars chan1 = constant {data_channel:
               -monitor_type DC_ALL,
               -monitor_demon lambda;
            error1:e1->error1:e0;
            error1:e2->error1:e1;
            chan1:value->error1:e2;
         servo_ks1:add(error1);
      printf(error1:e0, error1:e1, error1:e2, 'Joint 1
                               error = %, %, %, \n');
                        endlambda,
               -channel 1};


/*****************************************************/
/*                                                 */
/*      The interface from the robot simulator      */
/*      to the second servo expert                  */
/*                                                 */
/*****************************************************/

vars error2 = {error:
                 -name "error2",
                 -period 0.4};
vars servo_ks2 ={KS:};

vars chan2 = constant {data_channel:
               -monitor_type DC_ALL,
               -monitor_demon lambda;
            error2:e1->error2:e0;
            error2:e2->error2:e1;
            chan2:value->error2:e2;
         servo_ks2:add(error2);
      printf(error2:e0, error2:e1, error2:e2, 'Joint 2
                               error = %, %, %, \n');
                        endlambda,
               -channel 2};
```

```
/*******************************/
/*                             */
/*      THE ROBOT SIMULATOR    */
/*                             */
/*******************************/

#include <stdio.h>
#include <math.h>
#include "block_sockets.c"
#include "usocket.c"
float qd1, qd2, rdx, rdy, p=100.0;

main()
{
    float t, fi, rx, ry;
    float ei1=0.0, qdr1=0.0, q1;
    float qddr1, e1=0.05;
    float om1=1.0, ri1=0.0, taud1=0.01, delt=0.8;
    float dom1=0.0, dri1=0.0, dtaud1=0.0;
    float om1mx=2.0, ri1mx=0.1, taud1mx=1.0;
    float om1mn=0.5, ri1mn=0.0, taud1mn=0.0;
    float psen=5.0;
    float maxmin1,high1,low1;
    int n=500, edn=16, i=1, test=2;
    float distb1();
    float min(), max();

    float ei2=0.0, qdr2=0.0, q2;
    float qddr2, e2=0.05;
    float om2=1.5, ri2=0.0, taud2=0.01;
    float dom2=0.0, dri2=0.0, dtaud2=0.0;
    float om2mx=2.5, ri2mx=0.1, taud2mx=1.0;
    float om2mn=0.5, ri2mn=0.0, taud2mn=0.0;
    float maxmin2,high2,low2;
    float kinem(), distb2();

    FILE *fp, *fopen();

    fflush(stdout);
    muse_socket_to();

    rdx=1.0;
    rdy=0.25;
    kinem(0.0);
    q1 = qd1 - e1;
    q2 = qd2 - e2;

    maxmin1=e1;
    high1=e1;
    low1=e1;

    maxmin2 = e2;
    high2 = e2;
    low2 = e2;

    while (i <= n) {

    rx = 2.0 * cos(q1) + cos(q1 + q2);
    ry = 2.0 * sin(q1) + sin(q1 + q2);
    printf ("%8.4f\n", rx);
```

```
    printf ("%8.4f\n", ry);

    qddr1 = om1 * om1 * (e1 + ri1 * ei1 - taud1 * qdr1)
            + distb1(i);
    qdr1 = qdr1 + qddr1 * delt;
    q1 = q1 + qdr1 * delt;

    qddr2 = om2 * om2 * (e2 + ri2 * ei2 - taud2 * qdr2)
            + distb2(i);
    qdr2 = qdr2 + qddr2 * delt;
    q2 = q2 + qdr2 * delt;
(float) (fi = i);
    t = fi * delt;
    kinem(t);

    e1 = qd1 - q1;
    e2 = qd2 - q2;
    ei1 = ei1 + e1 * delt;
    ei2 = ei2 + e2 * delt;

/*  printf ("Joint error = "); */
/*  printf ("%8.4f\n", e1);    */
/*  printf ("%8.4f\n", e2);    */
/*  printf ("%8.4f\n", q1);    */
/*  printf ("%8.4f\n", q2);    */

    if (i % edn == 0) {
        if (maxmin1 == high1)
           maxmin1 = low1;
         else if (maxmin1 == low1)
           maxmin1 = high1;
         else if (test > 0)
           maxmin1 = high1;
         else
           maxmin1 = low1;

       sleep(2);
       muse_send_float(maxmin1, 1);
/*       printf ("%8.4f", maxmin1);    */

       fp = fopen("data1", "r");
       fscanf (fp, "%f%f%f", &dom1, &dri1, &dtaud1);
       fclose(fp);
/*       printf ("Joint 1 = %8.4f, %8.4f, %8.4f; ",
                  dom1, dri1, dtaud1); */

       om1 = om1 + dom1 * (om1mx - om1mn)/psen;
       ri1 = ri1 + dri1 * (ri1mx - ri1mn)/psen;
       taud1 = taud1 + dtaud1 * (taud1mx - taud1mn)/psen;

       if (om1 < om1mn)
          om1 = om1mn;
       if (om1 > om1mx)
          om1 = om1mx;
       if (ri1 < ri1mn)
          ri1 = ri1mn;
       if (ri1 > ri1mx)
          ri1 = ri1mx;
       if (taud1 < taud1mn)
          taud1 = taud1mn;
       if (taud1 > taud1mx)
          taud1 = taud1mx;
/*       printf ("Joint 1 PID = %8.4f, %8.4f, %8.4f\n",
                  om1, ri1, taud1); */

       if (maxmin2 == high2)
```

```
                maxmin2 = low2;
              else if (maxmin2 == low2)
                maxmin2 = high2;
              else if (test > 0)
                maxmin2 = high2;
              else
                maxmin2 = low2;

            sleep(2);
            muse_send_float(maxmin2, 2);
/*          printf ("%8.4f", maxmin2);     */

            fp = fopen("data2", "r");
            fscanf (fp, "%f%f%f", &dom2, &dri2, &dtaud2);
            fclose(fp);
/*          printf ("Joint 2 = %8.4f, %8.4f, %8.4f; ",
                    dom2, dri2, dtaud2); */

            om2 = om2 + dom2 * (om2mx - om2mn)/psen;
            ri2 = ri2 + dri2 * (ri2mx - ri2mn)/psen;
            taud2 = taud2 + dtaud2 * (taud2mx - taud2mn)/psen;

            if (om2 < om2mn)
              om2 = om2mn;
            if (om2 > om2mx)
              om2 = om2mx;
            if (ri2 < ri2mn)
              ri2 = ri2mn;
            if (ri2 > ri2mx)
              ri2 = ri2mx;
            if (taud2 < taud2mn)
              taud2 = taud2mn;
            if (taud2 > taud2mx)
              taud2 = taud2mx;
/*          printf ("Joint 2 PID = %8.4f, %8.4f, %8.4f\n",
                    om2, ri2, taud2);   */

            test = test * (-1);
            }

            high1 = max(maxmin1, e1);
            low1 = min(maxmin1, e1);

            high2 = max(maxmin2, e2);
            low2 = min(maxmin2, e2);

        ++i;
        }
            muse_close_socket();
}


float min(x, y)
float x, y;
{
    if (x < y)
        return (x);
    else
        return (y);
}


float max(x, y)
float x, y;
{
    if (x > y)
```

```
        return (x);
     else
         return (y);
}

float distbl(i)
int i;
{
if (i == 100)
   return (0.05);
if (i == 200)
   return (-0.05);
if (i == 300)
   return (0.1);
if (i == 425)
   return (-0.1);

   return (0.0);
}

float distb2(i)
int i;
{
if (i == 100)
   return (0.1);
if (i == 200)
   return (-0.1);
if (i == 300)
   return (0.05);
if (i == 425)
   return (-0.05);

   return (0.0);
}


/*****************************************/
/*                                       */
/*          Inverse Kinematics           */
/*                                       */
/*****************************************/

float kinem(t)
float t;
{
float trm, alph;

if (t < p) {
  rdx = 1.0;
  rdy = t * t /(2.0 * p * p) + 0.25;
};
if (t >= p && t < 2.0 * p)  {
  rdx = 1.0;
  rdy = (-1.0) * t * t /(2.0 * p * p) + 2.0 *
        t /p - 0.75;
};
if (t >= 2.0 * p && t < 3.0 * p) {
  rdy = 1.25;
  rdx = 0.5 * (t/p - 2.0) * (t/p - 2.0) + 1.0;
};
if (t >= 3.0 * p)    {
  rdy = 1.25;
  rdx = (-0.5) * (t/p - 2.0) * (t/p - 2.0) +
        2.0 * (t/p - 2.0);
};
```

```
trm = rdx * rdx + rdy * rdy;
qd2 = acos((trm - 5.0)/4.0);
alph = asin(sin(qd2)/sqrt(trm));
qd1 = atan(rdy/rdx) - alph;
return;
}

/*********************************************/
/*                                           */
/*  Inverse kinematics for the faster turn   */
/*                                           */
/*********************************************/

float kinem(t)
float t;
{
float trm, alph;

if (t < 1.5 * p) {
  rdx = 1.0;
  rdy = t * t /(3.0 * p * p) + 0.25;
};
if (t >= 1.5 * p && t < 2.0 * p)  {
  rdx = 1.0;
  rdy = (-1.0) * t * t /(p * p) + 4.0 * t /p - 2.75;
};
if (t >= 2.0 * p && t < 2.5 * p) {
  rdy = 1.25;
  rdx = (t/p - 2.0) * (t/p - 2.0) + 1.0;
};
if (t >= 2.5 * p)    {
  rdy = 1.25;
  rdx = (-1.0) * (t/p - 4.0) * (t/p - 4.0)/3.0 + 2.0;
};

trm = rdx * rdx + rdy * rdy;
qd2 = acos((trm - 5.0)/4.0);
alph = asin(sin(qd2)/sqrt(trm));
qd1 = atan(rdy/rdx) - alph;
return;
}
```

Lecture Notes in Control and Information Sciences

Edited by M. Thoma and A. Wyner

Lecture Notes in Control and Information Sciences

Edited by M. Thoma and A. Wyner

Lecture Notes in Control and Information Sciences

Edited by M. Thoma and A. Wyner

Vol. 117: K.J. Hunt
Stochastic Optimal Control Theory
with Application in Self-Tuning Control
X, 308 pages, 1989.

Vol. 118: L. Dai
Singular Control Systems
IX, 332 pages, 1989

Vol. 119: T. Başar, P. Bernhard
Differential Games and Applications
VII, 201 pages, 1989

Vol. 120: L. Trave, A. Titli. A. M. Tarras
Large Scale Systems:
Decentralization, Structure Constraints
and Fixed Modes
XIV, 384 pages, 1989

Vol. 121: A. Blaquière (Editor)
Modeling and Control of Systems
in Engineering, Quantum Mechanics,
Economics and Biosciences
Proceedings of the Bellman Continuum
Workshop 1988, June 13–14, Sophia Antipolis, France
XXVI, 519 pages, 1989

Vol. 122: J. Descusse, M. Fliess, A. Isidori,
D. Leborgne (Eds.)
New Trends in Nonlinear Control Theory
Proceedings of an International
Conference on Nonlinear Systems,
Nantes, France, June 13–17, 1988
VIII, 528 pages, 1989

Vol. 123: C. W. de Silva, A. G. J. MacFarlane
Knowledge-Based Control with
Application to Robots
X, 196 pages, 1989